To Audrey Coyle
from the 'author'
Bill Bristowe
(doesn't this sound pompous!)
Jan. 1975

A BOOK OF ISLANDS

A BOOK OF ISLANDS

W. S. Bristowe

LONDON: G. BELL & SONS

FIRST PUBLISHED 1969

PUBLISHED BY G. BELL & SONS LTD.
YORK HOUSE PORTUGAL STREET LONDON WC2

SBN 7135 1529 5

PRINTED IN GREAT BRITAIN BY
THE CAMELOT PRESS LTD. LONDON AND SOUTHAMPTON

TO MY FRIEND

THE LATE MAXWELL KNIGHT

WHO SUGGESTED THIS BOOK
SHOULD BE WRITTEN

CONTENTS

LIST OF ILLUSTRATIONS

Plates

BETWEEN PAGES 112 AND 113

Maps and figures

Note: Apart from the wheelhouse on South Uist and the bronze brooch by T. C. Lethbridge, the charmel by S. Ross-Craig and and the wolf spider by Arthur Smith, all the line drawings were made by Edgar Holloway, some from sketches by the author.

Preface

'EVERY seaman, every wanderer on the deep, has hearkened to the decoy of that ideal island, and where is the landsman with soul so dead . . . who in his homesick moments has never caught its enchanting echo?' Life and literature amply support Walter de la Mare's claim, for it would be easy to list rich and poor, tycoons and beachcombers, poets, artists and scholars, naturalists and children, cut-throats, pirates and evangelists who have dreamed of an island life. Fortunately they would not all settle for the same island.

Dreams unrealised contribute to happiness and should be nourished. If reality falls short of expectation, a fresh dream can be built. A small desert island could sustain its charm for an hour, a week or a month, according to different temperament, but bliss would remain for few who sought permanent marooning. Leave-taking before the period of enchantment has lapsed will incite a thirst for another island over the horizon, where hidden treasure, or fresh forms of life, or unrivalled beauty or even fairies might be found.

Juan Fernandez or the Galapagos may now be beyond my reach but imaginings can still provide a thrill.

The unknown and the unusual in beautiful settings have been my special spurs, but visits to well over a hundred islands have never been made with a book in mind. During a busy life opportunities to accompany expeditions, to spend holidays or to play truant on islands in the course of business trips have been seized. This will explain the uneven pattern of my island stories and the complete omission of some like Farne or the Isles of Scilly where my diaries were too full of natural history to be of general interest.

The islands range from the Arctic to the Equator and far beyond. They have provided discoveries, mysteries and dangers besides evil, beauty and peace. Although this is no guide book it contains forgotten history and myth, archaeology, natural history, and something of the ways of those who live on islands.

CHAPTER ONE

In the Arctic Mists of Jan Mayen

NO past acquaintances of mine such as Clarabutt, Hogsflesh or Innocent were so well suited to their names as Chaworth-Musters. Jim looked unusual, he was unusual, and he had an unusual shade of light mustard hair with a darker mustard suit to match.

We had both arrived at Cambridge in October 1920 and his long, lean uniquely square-shouldered figure had become a familiar feature as it drifted through the courts of Caius like some solitary bird. There was nothing unfriendly about his small blue eyes, his long protruding nose or his receding chin but he neither played games nor chose to mingle in the college activities.

We met by chance in our first term through finding ourselves sitting next to one another at dinner and this brought an invitation to coffee in his digs. Books and maps were strewn everywhere, but in a central position on his table was spread an old map of a curious island called Jan Mayen. His eyes lit up when he saw me examining it.

Jan Mayen, he explained, was some 300 miles north of Iceland and a similar distance from the east coast of Greenland. The island had seldom been visited since it ceased to be a whaling station long ago and never by anybody with scientific designs except an Austrian party in 1882/3 who had made this map and done some preliminary work on the natural history.

The map fascinated me. It showed an island shaped like a cutlet and thirty-five miles long, the meaty end being taken up entirely by a huge glaciered mountain of 8,350 feet. The long narrow tail was toothed with lesser peaks and there was a lagoon on each side of the low central waist. Features of the map commemorated walrus and bear as well as seven Dutchmen who had perished from scurvy many years earlier.

Jim, it transpired, was determined to get to Jan Mayen during the following summer if he could find a few companions to share the expense. The challenge of the unscaled Mt. Beerenberg and the study of wild flowers were his special spurs.

Being a naturalist myself, and having had an interest in desert islands ever since the childhood reading of adventures like *Coral Island* and *Robinson Crusoe*, I was entranced by the thought, but a big gulf was fixed between dreams and reality. Life so far had been sheltered with never a trip abroad or even a hill to climb higher than a Dorset down. A project like this scared me and I would be intensely reluctant to ask a generous but none too prosperous father for funds.

Thoughts of that exciting map left me in no peace so a visit was paid to my schoolfriend, Tom Lethbridge, at Trinity whose toughness, experience and self assurance inspired me with confidence in any emergency. It was Tom who had awakened an interest in archaeology at school while his brilliant paintings of insects had provoked my envy. Together we trooped round to Jim's lair and suddenly all doubts were resolved and we both decided to go with him and a friend whom he was shortly to entice into the team.

News of our intentions reached the ears of James Wordie, a tutor at St. Johns, in the Easter Term and he said he would like to come with us. It was a symptom of our youthfulness at nineteen that we hesitated lest his extra burden of fourteen years would be too much for him! Wordie was a geologist with experience of mountain climbing who had been with Shackleton in the Antarctic and with W. S. Bruce in Spitsbergen so he was obviously just the man we needed.

Although it was inevitable that Wordie should become the leader, the planning was delegated to Jim and he returned from Norway with the major obstacle of transport neatly arranged. He had found that a Norwegian engineer called Hagbard Ekerold was leaving early in July to erect a weather station on Jan Mayen and that he was willing to provide passages for a modest fee.

Stepping ashore in Bergen on 10th July in elderly tweeds we were confronted by a giant of a man in frock coat and top hat. This was Ekerold. He was 6 feet 4 inches in height, swarthy and good-looking in a fierce way with aquiline features, a strong chin and penetrating brown eyes.

Two pieces of disappointing news greeted us. The first, that one of his two boats was not yet ready to sail. The second, that he had promised accommodation to a Professor P. L. Mercanton of Lausanne whose burning ambition it was to climb Beerenberg. The only consolation we could find lay in Mercanton's reputation as a skilled mountaineer and glaciologist.

The professor turned out to be a portly middle-aged man with a bridged nose and limpid hazel eyes whose cheerful disposition inspired him to sing patriotic songs whenever the sun shone. We could scarcely help liking him although he was so helpless and so aggravating that he became the constant butt for jokes and criticism, all of which he took cheerfully and without malice.

Looking back on Wordie with hindsight after many years of subsequent friendship, he still remains an enigma. He was slightly built with a thin face from which peered eyes made beady by rimless glasses. He loved to make jokes in his broad Scottish accent and, with mouth turned down at the corners, he would tell us such mixtures of fact and fantasy that almost every utterance was greeted with amused scepticism. Wordie was a splendid companion but neither by presence nor personality was he a born leader, though he usually got his way by some circuitous route rather than by direct order or request. None of us would have predicted his later successes as Master of St. Johns, Vice Chancellor of Cambridge, President of the Royal Geographical Society and the award of a knighthood, although he deserved them all. Shrewd common sense rather than brilliance was his great asset.

Wordie was a mine of information about Arctic and Antarctic exploration so we extracted all he knew about Jan Mayen during the period of waiting in Bergen.

Jan Mayen was known to the early Norwegian explorers who colonised Greenland in the 10th to 14th centuries and they were thought to have referred to it as Svalberg in their sagas. The decline of Norway as a seafaring nation led to the island being forgotten for the next 300 years.

It was rediscovered by a host of mariners early in the 17th century but each kept his discovery dark in order to avoid competition from others in search of whales. The name which has stuck derives from a Dutch captain, Jan May, who found it in 1614 and this name superseded that of Joris Eylandt given to it on their chart by Joris Carolus, his mate, in self glorification.

There is evidence that English whalers had known it earlier but secrecy had prevented their private names gaining priority. *Hudson's Tutches* almost certainly referred to Jan Mayen from Hudson's sight of it in 1607 and Hull whalers seem to have been calling it Trinity Island by 1611. The first fully authenticated English visit and excellent description was by Robert Fotherby in 1615 when he called it Sir Thomas Smith's Island.

Other 'discoverers' were soon to call it Young's Foreland, Mauritius, Pico and the Isle de Richelieu, the last name being due to a Biscayan whaler, Jean Vrolicq, who claimed to have found it in 1612.

The Dutch erected vats for rendering down the blubber on the island as early as 1616 and it was due to this Jean Vrolicq pillaging their whaling depot in 1632 that the seven Dutchmen met their fate through staying on guard during the following winter.

Ruthless destruction of whales continued throughout the 17th and 18th centuries until they became so scarce that Jan Mayen was abandoned and again sank into obscurity. During the past hundred years and more no visitors had spent more than a day or two except for occasional fox trappers and the Austrian Expedition of 1882/3.

As to volcanic activity, Wordie had traced only two reports of eruptions. Both had been insignificant. Jacob Laab reported flames and billowing black smoke from a small crater near the base of Mt. Beerenberg in 1732 and Scoresby another small eruption from about the same spot in 1818.

Returning to our own plight and the delays now caused by government regulations and restrictions in Bergen, Ekerold at last managed to escape with both boats to Aalesund and thence to Bratvaag. In this way further detention and the serious risk of his having to jettison some cargo or passengers was circumvented. Thanks to this ruse both boats finally sailed from Bratvaag at 09.00 hours on 1st August minus the skipper of one of them who had taken fright at their grossly overloaded condition. There was also something about foul play and the death of a seaman, the full details of which never reached us.

Three important weeks of Arctic summer had been lost but we were thankful to be on our way at last. The fear of being chased and arrested for having sailed without clearance papers still remained, but spirits lifted with every sea mile covered.

Both boats were remarkably small for what they carried. *Isfuglen*'s vital statistics were 70 feet in length and 45 tons, *Polarfront* a mere 45 feet and 25 tons. Twelve men sailed in *Isfuglen*, including our party of five, while the nine in *Polarfront* included Ekerold and Mercanton. Cargo filled every cranny and decks were piled high with massive wireless masts, wooden sections for huts, heavy machinery, vast oil drums and packing cases.

The sea was calm and the sun shone as Norway's snow-streaked blue mountains faded into the distance. Then, just as all our anxieties had vanished, *Polarfront*'s engine stopped throbbing and Mercanton's song across the smooth water was stilled in his throat. Although the engine's fault was soon mended their captain now discovered her compass was out of order which augured ill should the boats become separated, so she suffered the indignity of a tow rope from *Isfuglen*.

These broad-beamed boats were rounded like tubs to combat an ice squeeze so they rolled incessantly during the next two days of rough weather. Whatever the state of our health we stayed on deck during the daytime and only retired to the dark damp forecastle at night where two had to share each single bunk. The sea sweeping over the gunwales, the sound of rushing water and the fulmar petrels skimming the waves beside us were infinitely preferable to the smell of mildew and seal blubber in that fore-castle.

Calm weather on the fourth day was followed by high winds and heavy seas on the fifth which made us climb higher up the recumbent wireless masts to avoid the curtains of green water which swept the decks continuously. Now in the teeth of half a gale it was *Isfuglen*'s turn to falter and she wallowed deeply as waves crashed down on her, sending darts of ice-cold water in our faces. The inadequacy of auxiliary sails without the engine's boost was only too apparent but after three anxious hours the engine coughed and spluttered into action again.

The sixth day was calmer and much colder. The presence of kittiwakes and skuas confirmed our hopes that we were drawing near to land but thick mist made it obvious that the glorious view of Beerenberg for which we had hoped would be denied us. Indeed horrid thoughts of sailing past so small a target as Jan Mayen in this muslin cloud had begun to assail us when suddenly white waters breaking on a dark coastline only a few miles away loomed out of the mist at 07.00 hours. This was a moment of intense excitement.

Snowy slopes and a glacier became dimly visible and then, for an instant, the white summit of Beerenberg as we made a westward approach to Jameson Bay. It was a beautiful mountain with symmetrical slopes steepening to its wide crater rim but somehow it was less gigantic than our imaginations had pictured for what was believed to be the highest mountain in the Arctic.

At this season there was still permanent daylight so landmarks could be picked out from the map. The left side of Jameson Bay was bounded by Egg Bluff, the remnant of an old ash crater from which thin wisps of steam were rising. A black sandy beach, strewn with hefty pieces of bleached driftwood, lay ahead stretching back to rugged hills of red and dark grey covered in places with yellow moss. To the right of the beach a jagged grey lava flow led to the foothills below the mantle of glaciers and snow which draped the shoulders of Mt. Beerenberg.

The boats were anchored and not a moment was wasted in rowing to the shore. The thrill of landing on our desert island made us want to skip with joy but the inclination was defeated by finding our legs had sunk calf-deep into fine black sand. Bare-

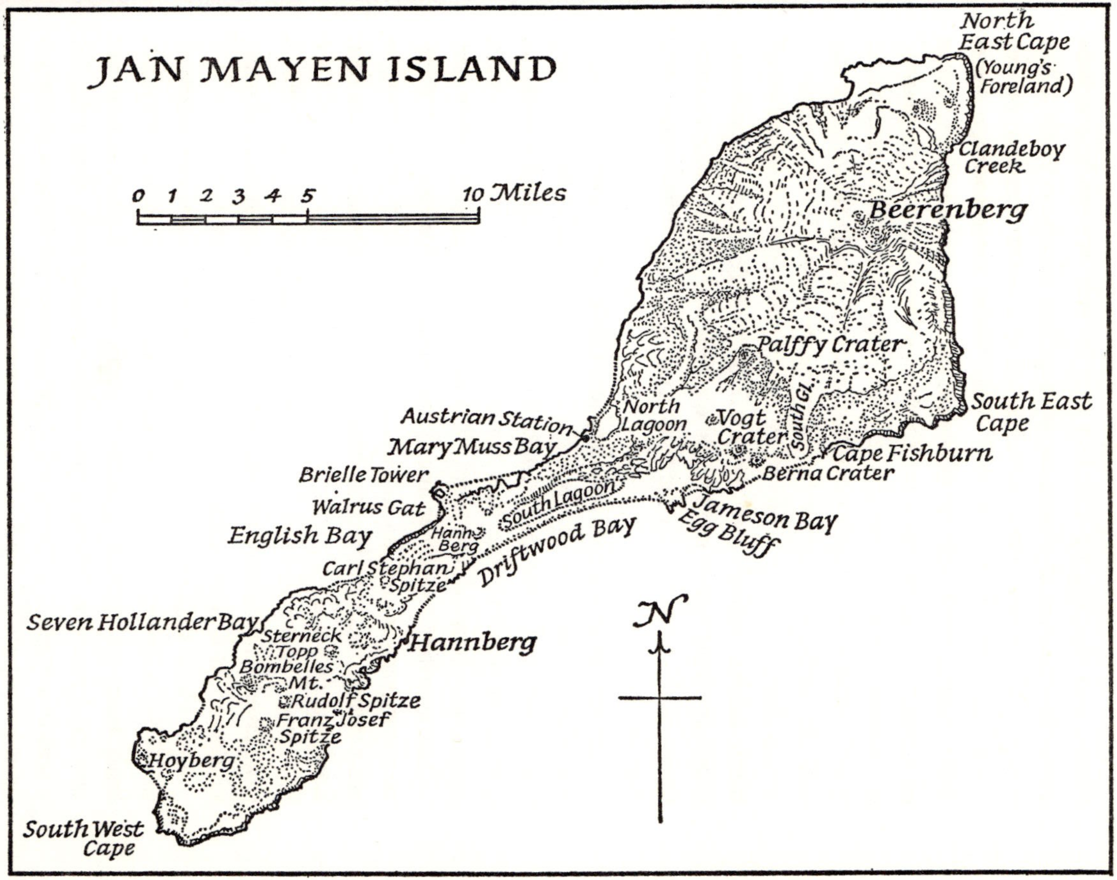
JAN MAYEN ISLAND
0 1 2 3 4 5 10 Miles
North East Cape (Young's Foreland)
Clandeboy Creek
Beerenberg
Palffy Crater
South East Cape
North Lagoon
Vogt Crater
South Gl.
Cape Fishburn
Berna Crater
Austrian Station
Mary Muss Bay
Brielle Tower
Walrus Gat
South Lagoon
Jameson Bay
Egg Bluff
English Bay
Hann Berg
Driftwood Bay
Carl Stephan Spitze
Seven Hollander Bay
Sterneck Topp
Hannberg
Bombelles Mt.
Rudolf Spitze
Franz Josef Spitze
Hoyberg
South West Cape
N

footed and tugging up our feet step by step we made our way across the black sand glistening with yellow olivine crystals to the nearest hill-slope. Here patches of old snow nestled in the sheltered crannies of loose red rocks while the smoother slopes were matted with yellow moss and green-grey lichen. Clumps of beautiful rock plants were thickly grouped as though planted by human hand. Cushions of *Saxifraga oppositifolia* and *Silene acaulis* were ablaze with pink flowers and beside them were the delicate white blooms of *Ranunculus glacialis*. The turning of stones disclosed insect life in quantity and two kinds of small black spider against one recorded by the Austrian expedition. Our happiness was complete as we returned to the boat at two-thirty in the morning in the pink glow of a combined sunset and sunrise.

The real introduction to the island came next day. Ekerold had told us that four Norwegian fox trappers had been spending the winter at the old Austrian station across the island's waist, so it was decided that the first walk should include a call on them despite the disdainful warning that they would be covered in lice.

The weather was clear and gave unimpeded views of Beerenberg's snowy slopes, its massive beauty and stature increasing as we mounted the neck crest at an altitude of perhaps 1,000 feet. Looking down on the northern side of the island we could see hill-slopes descending to a large blue lagoon enclosed by a castellated hill and narrow sand-bar with a white building of some kind near its western end.

The clean-shaven but otherwise unkempt-looking trappers provided a greeting with hot mugs of coffee inside a cosy hut built from part of an upturned ship's hull. The Austrian hut was used to store their 120 fox skins, a poor catch, they said, for a winter's work. Their relief ship was overdue and, with only a month's supply of food left, they were getting anxious.

A prospector called Birger Jacobsen was spending the summer in a hut at the other end of the lagoon with two companions, and a course was next directed to his encampment The clear weedless water of the lagoon was fresh to the taste and some soundings, taken later from Jacobsen's boat, showed that its depth exceeded six fathoms. Small trout parr came as a surprise though crustacea, algae and diatoms were also present to provide them with food.

Jacobsen was a plump man with long dark hair and protruding blue eyes who told us in perfect English that no minerals of any value existed on Jan Mayen. The only thing of any interest was the now headless figurehead of *Ariadne* left by Lord Dufferin during a few hours' visit in 1856.

Presently Jacobsen's younger companion started to strum on a mandolin and sing Norwegian sea shanties which naturally aroused Mercanton to respond with Swiss songs. Greater shyness and less melody was provided by our rendering of English and Scottish songs.

Mist was forming as we set out on the homeward journey. A backward glance from the hill crest gave a weirdly beautiful view of grey hill slopes, slashed with streaks of bright yellow moss, stretching down to the lagoon partly enclosed by the arm of a black castellated ridge strongly resembling a hill fort.

The second day was spent in getting the gear ashore and setting up camp in Jameson Bay close to a lava flow where the sand was reasonably firm. Fresh water could be had either by digging holes or by melting snow from near-by patches. For cooking there was a choice between primus stoves or a fire stoked with driftwood ranging from trunks to branches which had made their long journey from Siberian rivers in the westerly Arctic current.

The bell tent was large enough for all the sleeping-bags including that of Mercanton whom we had adopted as a delightful though exasperating companion. The stores were stacked near by and it took only two days to realise that one of us had to stay on guard throughout the day against the pilferings of a thirsty trapper and a hungry blue fox. A turn in camp every sixth day was never wasted as there were always specimens to sort, notes to write and sketches to finish. Then, at odd moments, there was the attentive blue fox to play with. After trotting to within six feet he waited expectantly to be fed or chased and seemed to enjoy both equally well. When the chaser grew tired he would follow at his heels back to the camp and almost feed from his hand.

The party thrived on the singularly boring diet consisting of little except corned beef, sardines, tinned fruit, porridge, marmalade, butter and mouldy biscuits contaminated by sea water. There were also limited supplies of pemmican which made delicious soup and some concentrated form of nougat invented for Sir Ernest Shackleton in the Antarctic. Fortunately some of us had guns and found we preferred Brünnich guillemots, puffins and glaucous gulls to the other birds which were sampled, while there were gala days feasting on black steaks from a bearded seal.

As for Mercanton, he must have been a fatalist to come on such an expedition without any provisions of his own, without a gun and quite unteachable even as an assistant cook or washer-up. We accepted him as we found him and even laughed when it was discovered that after enjoying breakfast with us he called on Ekerold each morning for a second one.

Walking long distances across soft sand or loose rocky hill slopes was so laborious that it was decided to establish two minor bases, at Hann Berg to the south and on the lower slopes of Beerenberg at 2,700 feet. The small tents could accommodate three people and the first would help parties to explore the south end while the second would enable the climbing party to spend a night on the mountain.

The climbing party was limited by the size of this tent so only one of the original team could be selected as Mercanton and Wordie were obvious members. In view of this it was decided that a second attempt would be made later, whether the first had been successful or not. Tom was rightly selected for the first climb though it was disappointing that Jim could not be included.

Jan Mayen had proved to be an island of mist so it was necessary to wait for clear weather in which to make the assault on Beerenberg. Meanwhile daily excursions were made into the hills, Tom's interest in geology usually leading him to accompany Wordie while Jim's and my natural history quests kept us together.

The most rewarding direction for the naturalists lay along the south-east coast. Here the lava cliffs below Beerenberg provided nesting places for countless puffins, Brünnich guillemots and little auks. Beyond was a beach where turnstones, sanderlings and ringed plover glutted themselves on fleshy molluscs and pink-eyed sandhoppers. An occasional snow bunting winged from rock to rock near a torrent of water flowing down the mountain-side from one of its glaciers.

High black cliffs barred progress beyond Cape Fishburn but the debris at their base was well vegetated with plants including the small purple *Veronica alpina* and a tiny black spider (*Meioneta nigripes*) not seen elsewhere.

The coastal strip, and this one in particular, had most of the fauna and flora clustered together. Some plants and insects struggled up the hills but their slopes were mostly bare or covered in moss and lichen. The only 'tree' was the four-inch Least Willow (*Salix herbacea*). Many of the pretty wild flowers were so reminiscent of rock gardens that one instinctively thought of weeding the common dandelion which looked incongruous amongst them. A favourite was the beautiful glaucous-leafed Oyster Plant (*Mertensia maritima*) whose vivid blue flowers gleamed from the rocky beaches above high tide level.

10th August was the first day on which the whole island was completely free of mist and cloud so we all assisted the climbing party to their advance camp with their gear. Mercanton's instruments of science seemed to be particularly heavy and it was only discovered after arrival that the burden had included three quart bottles of a mineral water which he had promised to take up the mountain in return for the makers' financial assistance.

The route from the base camp lay up a dry valley between two little rounded craters called Esk and Vogt. Tell-tale black crystals and volcanic bombs helped to confirm that it was the Vogt crater which had been seen in eruption in 1818.

Glassy lava took the place of crumbly rocks further up the valley until another older and eroded crater was reached at about 2,000 feet. Above this came some frontal moraine, or rock debris carried from the higher slopes by glaciers, and finally the advance camp at 2,700 feet after crossing short patches of glistening snow. Even in this inhospitable spot a few brave *Ranunculus glacialis, Saxifraga caespitosa* and a *Draba* were flowering amongst the rocks, plant midges were in the air and springtails were pursued by tiny spiders in the moss and lichen.

The bearers returned to the base camp knowing that the assault party would set out early next morning. When that time came they saw with anxiety that cloud had descended on the mountain. As the day advanced this cloud thinned but never gave more than occasional views of the summit.

The intended timetable catered for a second night at the advance camp so we woke early on the 12th and kept our eyes skinned all the morning for any sign of them. During lunch we decided that they must have deferred the climb on account of the previous day's cloud and would not now return until the morrow, so I remained in camp while the others went out to celebrate the 12th by shooting birds for the pot. Soon after they had gone I saw Tom approaching. He was alone so I feared the worst and hurried out to meet him! It turned out that he had come on ahead of the other two who were taking observations.

They had climbed Mt. Beerenberg! That was the essential news. Later we were all to get such detailed accounts from Tom and Wordie, and at least three from Mercanton, that we almost felt we had been with them after balancing Tom's understatements with the epic descriptions of Mercanton.

At 5,600 feet they had come to a cairn marking the Austrian expedition's highest survey point. So far the going had been easy up a gentle 1 in 7 ice slope but here it was necessary to work out the best route to the dimly visible summit. After brief discussion they decided to make their way along a sheltered snow-covered depression on the southern side of what might be a slippery ice ridge. A number of small crevasses were encountered but these were all bridged with firm snow until they reached about 7,200 feet. Here there was a gaping crevasse with no sign of any route round it, so each in turn had to jump six feet in an upward direction from a snow take-off.

No other difficulty was experienced until their further passage was blocked by a vertical ice wall at 7,600 feet. It was here that

Mercanton's skill and experience came to the rescue. Steps were cut with an ice-axe and after surmounting this final obstacle the crater rim was reached up a steep snow slope.

Through the mist they could now look down into a great ice-filled crater 500 to 800 feet deep and about half a mile across. Through a breach in its northern wall could be seen a great glacier pouring in a series of massive ice-falls right down to the sea.

The party had not yet reached the highest point so they moved precariously in a westerly direction along the narrow rim. Mercanton was allowed the privilege of being the first person ever to set foot on the mound which marked the supreme summit and he became almost delirious with excitement. 'Ah! Ah! Ah!' he shouted, 'I Mercanton, I have seduced ze Beerenberg. She is no longer a virgin.' And then, as though expecting Vulcan to erupt with fury, 'What will her father say?' Singing followed and the opening of his mineral water bottle as though it were the best champagne.

The climb had taken eight hours from the advance camp and now the party busied themselves taking photographs, making drawings, hoisting Swiss and Union Jack flags, testing the altitude by various methods and burying a metal cylinder as proof of conquest. Their estimate of the mountain's height was 8,090 feet compared with the Austrian figure of 8,350, but it has since been reckoned to be under 8,000 feet.

The ceremony of burying souvenirs in the metal cylinder did not pass without some amusement. Having inserted a piece of paper with the date (11th August 1921) and their signatures, Mercanton added part of a Swiss newspaper, a 5-centime piece and some other small unidentified object. Pressed for information, he explained, 'It is my little foolishness. It is the first tooth of my eldest daughter, Maya.' Tom's own little foolishness had gone unobserved. He had inserted a fly button so he was scarcely justified in laughing at Mercanton's more worthy souvenir, but then Tom is one of those stiff-upper-lip Englishmen who would rather die than admit their high degree of sentimentality!

Now the big objective had been accomplished, Jim and I were free to explore and make collections in the southern part of the island. With packs on our backs we trudged six miles across yielding sand to the small tent at Hann Berg. Part of the journey lay beside the South Lagoon, which was merely a shallow expanse of standing water, the only landmark of consequence being a high black finger of rock on whose ledges a quantity of kittiwakes were nesting.

Mist had turned into rain before camp was reached and all the driftwood was too sodden to use for a fire by which to dry our clothes, so we turned in early and hoped for better things in the morning.

The rain had stopped and we soon grew warm climbing a near-by valley to a plateau of moss-covered boulders. Leaping from one to the next like a couple of mountain goats we slithered down a loose cindery slope into a bay on the northern shore beside Walrus Gat, which had been one of the intriguing features on Jim's map in Cambridge. The wide bay of black sand was closed to the north by an impressive hill called Brielle Tower, which was turretted like some grand castle, while to the south were the dark perpendicular cliffs of English Bay. Bleached piles of whalebones were reminders of Jan Mayen's importance 200 years earlier and the remains of a large wooden hut commemorated the temporary home of the whalers themselves. Massive piles of driftwood covered the shore with numerous green glass floats which the sea had brought from fishing vessels. Out to sea swam quantities of eider duck, peaceful and unafraid.

The next day had the best weather of any experienced on Jan Mayen. This time a more southerly crossing of the island had been planned and the sun was so hot that we stripped to the waist before reaching a pass between snowy mountain peaks. The view of Beerenberg some twenty-five miles away was reminiscent of pictures of Fujiyama. Close at hand were the snowy slopes of Sterneck Topp and a strange Christmas pudding of a hill called Bombelles sprinkled with snow on its summit with numerous snow-filled furrows in its black cindery sides (Pl. 1).

At last we were standing in Seven Hollander Bay beside a wooden cross erected in memory of the seven Dutchmen who had died during the winter of 1633/4. Jim had read translations of their diary and reconstructed what he could remember.

They had watched their friends sail for home with heavy hearts in August 1633 and described their first walk during which they had seen nothing 'worth their observation'. They 'met with nothing'.

Gloom filled every page of the diary and a state of mental prostration overcame them as ice and permanent darkness hemmed them in during the winter months, making them easy preys to scurvy.

In mid-December they shot a bear and in late February they had glimpses of the sun above the horizon, but by March the effects of scurvy were well advanced.

We pictured them stumbling and crawling painfully up the hills in vain hope of seeing a ship's sails until even the fittest of them could manage the journey no longer. By the end of March their legs were 'scarce able to bear' them. On 3rd April were killed 'the only two pullets we had left'. Deaths soon followed and by 23rd April 'we were by this time reduced to a very deplorable state, there being none of us all except myself that were able to help themselves, much less one another, so that the whole burden lays upon me—and I perform my duty as well as I am able, as long as God pleases to give me strength'.

A week later the writer's failing hand could do no more than trace a few indecipherable words on the page.

The only consoling feature in this grim story was the loving care with which the sick man had tried to look after the invalids in a worse state than himself.

Our places were taken by Wordie and Tom on the following day so we returned to the base camp in time for Ekerold's party celebrating the completion of the living quarters for his wintering party. After midnight we were peacefully asleep when a commotion inside the tent awoke us. The engineer was tripping over our feet and yelling for whisky. He was obviously drunk and not another word could be understood so efforts were made to get him away, first by persuasion, then by gently linking an arm in his, and finally by throwing him out. Back he came with glaring eyes and fists raised, but just as a fight seemed inevitable the English-speaking cook arrived to re-establish peace.

The crew, it seemed, had fastened the dinghies insecurely when coming ashore and had just found them floating out to sea in the rising tide. Some rough raft of driftwood had been constructed on which the engineer had started to paddle in pursuit but this had capsized leaving him to swim the last few yards to the boats. Now he was freezing and believed he would die unless he was given whisky, but all were too frightened of Ekerold's wrath to seek help from him.

We dug up our only bottle of whisky and friendly relations were restored.

The steam coming from Egg Bluff was the only evidence of volcanic life so we clambered up the steep ashy slopes to make an examination. Here two small fumaroles were seen from which steam was rising and it was found that these were hot enough to boil a couple of deserted guillemots' eggs within a few minutes.

Returning from this little jaunt, we found a different type of fumarole in the shape of an erupting Ekerold. A change in wind during the previous day had made anchorage on the other side of the island necessary, but not before the *Polarfront* had nearly been wrecked on the beach. This bad seamanship had enraged Ekerold, but now the men had gone on strike after arriving half-way through the morning and being refused breakfast before noon.

Wordie's calm words and advice appeased Ekerold and he decided to bide his time until they returned for their lunch. Sure enough a disconsolate group was seen returning to work in less than an hour. From this time onwards they worked with increased vigour in the knowledge, no doubt, that the wireless station had to be completed before they could get back to their homes in Norway and none wanted to risk delays until the ice closed in on them.

Frost at night and snow on the hills heralded the end of summer conditions so a fine day on 29th August was selected for the second climb. The hope was held that this could be made in one day leaving the possibility of one night in the advance camp on the return journey, but this was destined not to be followed.

Jim seemed to have lost all interest (and I used to wonder in later years if he had received warnings of the heart weakness which was to cause his death several years later) so I set off with Wordie and Mercanton. This trip was an aggravating failure from start to finish. Departure was delayed by two hours because Mercanton had disappeared for his second breakfast with Ekerold and had had to mend his hypsometer after his return. The weather worsened as we climbed the rind-covered rocks, and snow now lay where the ground had previously been bare. The tent had collapsed and was frozen to the ground, so it became necessary to thaw knots in our mouths and re-erect it in an ice-cold wind. An uncomfortable night was then spent huddled together for warmth, while snow fell and the wind whistled round the tent.

Early next morning the weather had cleared and the summit looked enticingly near, but a strong icy blast was blowing down the slope into our faces. 'It is impossible,' said Mercanton and his words were echoed by Wordie. 'Besides,' added Mercanton, 'I have lost one of my gloves.'

This was the only occasion on which I got angry with Mercanton. Having reminded him of his promises, the delay in our departure time and his carelessness over his glove, I even picked up his ice-axe and started on my own. The wind numbed my face and penetrated my unsuitable woollen clothing, so I soon turned back knowing they were right.

Summer was at an end. Even the base camp had had snow and a sharp frost. People spoke of ice blinks in the sky—light reflected from ice-floes—and Ekerold's progress had been such as to enable him now to place the *Polarfront* at our disposal for the return voyage, provided we could manage it ourselves with one engineer and the cook who was useless as a manual worker.

The next two days were spent packing, loading our own equipment and Ekerold's superfluous gear, and saying our goodbyes to an island for which we had gained considerable affection despite its misty climate. Jim and I walked over to bid farewell to the trappers with a load of food without fully realising the drama which had begun to develop between them and Ekerold.

The trappers were near the end of their food supply and were thoroughly worried lest the relief ship's delay would cause them to spend another winter on the island. Ekerold was racing against time to complete the station before ice-flows made his departure necessary. It was due to this tense situation that he had developed a fierce hatred for the trappers who had been caught pilfering his stores and been sent away with a warning that they would be shot if they were seen in Jameson Bay again.

Later we were to hear the sequel. Ekerold spotted the four trappers approaching with shot guns, so he fired a warning shot over their heads and took cover behind a lava flow. For a time they threatened to return his fire unless he gave them food, but peace was declared on Ekerold's terms after a short parley. They dropped their guns and advanced on hands and knees while the guns were collected. It was then agreed that they would work for him in return for food and, if necessary, be given a passage back to Norway. He made them work hard but he certainly saved their lives.

The station was completed in time to send weather messages to Europe before the end of September while Ekerold in *Isfuglen* was dodging ice-floes on her homeward journey.

The *Polarfront* sailed at 20.30 hours on the 3rd September and gradually our island faded into the distance but we were still to catch a last glimpse of Beerenberg from a distance of eighty miles before her faint outlines merged with the haze.

Pairs of us now took watches in turn during which the ship was steered to an agreed course, sails were adjusted whenever necessary, distance readings were read from the ship's log and the position determined by sextant at noon. *Polarfront*'s rate of progress varied from marching speed to that of a fast walker, say 3½ to 6 knots, which meant that she usually covered not much more than a hundred miles in every twenty-four hours.

Things went almost too smoothly for the amateur crew and just as a sharp watch was being kept for the first sight of Norway on the sixth day the engine caught fire. Billowing smoke rose from the engine room, but fortunately this was extinguished and *Polarfront* limped on at reduced speed. Land was sighted at 08.15 hours and we were at anchor in Aalesund bay by 21.00 hours.

The adventure was over and Beerenberg had been climbed. The fun had been enough in itself so any additions to knowledge came as an extra bonus. The geologists had collected evidence pointing to the probable birth of Jan Mayen from vast volcanic upheavals starting about one million years ago. The naturalists had added many species to the known fauna and flora from amongst their collections of fifty-two insects and spiders and forty-three flowering plants and grasses (besides some fungi, lichens and mosses).

The younger members of the party had all gained something intangible which combined a feeling of self-reliance with a thirst for more adventures.

CHAPTER TWO

To an Enchanted Scottish Island

TOM LETHBRIDGE and I have had a rich variety of adventures together since Jan Mayen, sailing among Scottish islands, climbing mountains, exploring the islands of Eire, excavating Saxon cemeteries in the Cambridgeshire fens and even enquiring into witchcraft in the same region. None has been so remarkable or so inexplicable as that which befell us on a Hebridean island which did not like being visited.

So clearly do I remember the events on that Friday morning, 21st July, 1922, that my diary can contribute little I might have forgotten. Such were our feelings of mystification and awe that for years we kept silent, for fear of being mocked, until more recent knowledge of Scottish folklore had brought fresh significance to the events. Some may still mock but I can only describe what happened.

Plans for an expedition to Greenland with Wordie had fallen through so the three of us had decided to sail among the western islands of Scotland. The *Elspeth*, a sixteen-ton yawl, was chartered along with a skipper from Skye called John Robertson, a gnarled old Highlander who knew every treacherous tide and shoal in these western waters. The party was completed by Geoffrey Walford, an undergraduate friend, and Harker, an eminent old Cambridge geologist.

Geoffrey would have been an asset on any holiday but Harker was less digestible. He was a lonely bachelor on whom Wordie had taken pity and his musty old Norfolk jacket, with tight threadbare knickerbockers to match, must have served him well since his undergraduate days some forty years earlier. His drooping moustache and Petersen pipe gave his face a doleful expression and never once did we see his thin lined face lit by a smile. Nevertheless, as the weeks passed, we felt glad he had come. As a young man he had taken part in the geological survey of western Scotland and now for the first time he was revisiting the scenes of beauty where his career as a geologist had started. Although he had lost the art of showing his feelings we somehow knew he was bubbling inside with intense enjoyment.

From Gourock *Elspeth* sailed through the Kyles of Bute, the Crinan Canal and the tide races in Scarba Sound to the Isles of the Sea (Garvellochs) where rabbits drummed at our inspection of St. Brandon's humble 6th-century chapel, now lying forlorn among primroses and buttercups.

Sweet breezes bore us thence to Oban and to Tobermory where one of the intermittent hunts for Spanish treasure was in progress.

A golden eagle from Mull soared overhead as *Elspeth* made her way to the rocky sgurr of Eigg. Here a brief pause enabled a visit to be made to the historic cave in which the avenging MacLeods of Skye had smoked to death over 300 men, women and children of the Clan MacDonald in 1577. Wordie's contribution to history, that a MacDonald scout had given their hiding place away by artfully walking backwards in the snow, almost brought a smile to Harker's lips.

Anchoring that night in smooth waters at Rum we enjoyed one of those memorable sunsets which only a combination of sea, clouds and mountains can provide. The deep blue Cuillins of Skye, mirrored in the still sea, gradually changed to pink whilst the sky passed from pink to fiery crimson before the scene faded and was lost in grey shadow.

Rum possesses mountain grandeur, but to me it seemed to have neither the beauty of Skye nor the emerald brightness of low lying Canna. This feeling of near depression followed us up Halival (2,367 ft.) where we were pursued by Sir George Bullough's factor to tell us we were trespassing. How ludicrous this seemed, but all ended well with ample hospitality during the homeward journey owing to Sir George's brother being a bursar at Cambridge.

It was on this island in later years that botanists had the

temporary excitement of learning that plants unknown to Britain were growing there. But disillusionment was to follow. Alas, one amongst them had carefully planted these unique species.

Onwards we sailed to Loch Nevis, Mallaig and Broadford where the summit of the Old Woman (Bein-na-Caillich) gave such glorious close views of the Cuillins as to make us decide to visit them during the homeward journey.

After brief stays at Loch Torridon, Gairloch, the Summer Isles and Lochinver, whose phallic-shaped Suilven (2,399 ft.) had to be climbed, the Minch was crossed to Lewis and Harris.

One of the joys of this cruise had been that there were no set plans, so in crossing from Tarbert (Harris) to Skye a close approach to the Shiant Islands on Friday, 21st July, caused a change of mind.

The Minch had been docile and the legendary disturbance of these waters by the Blue Men of Gaelic mythology had hardly caused *Elspeth* so much as a bubble under her keel, so when she approached the lustrous green-capped islands of Shiant we decided to go ashore. After the dour greyness of Harris they lured us to closer inspection! *Elspeth* soon sailed into the wide bay formed by Garbh Eilean and Eilean Tighe where the third island

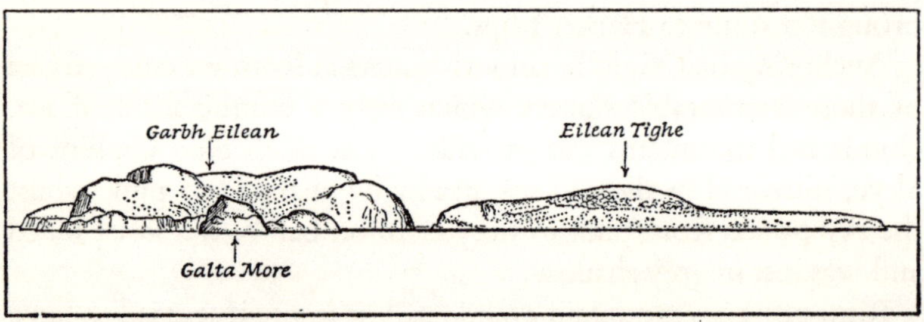

The Shiants

of Eilean Mhuire formed a shield to still the waters. Nowhere had we seen so many birds and the glassy waters were dotted with countless thousands of puffins, all so well fed as to show reluctance to move as we came amongst them. Guillemots and razorbills mingled with the puffins, and kittiwakes flew with their white grace around the boat.

A square-roofed gateway through Garbh Eilean's northern ramparts framed like a picture the distant view of Lewis. Sheer basalt columns, higher though less regular than on Staffa, were warmed by golden lichen reflected in the green sea. Screes cloaked with grass matched the emerald green of the island's smooth caps.

Eilean Tighe, smaller and lower, was connected to Garbh Eilean by a causeway of rounded beach stones. It gave the appearance of being little more than one small hill of brilliant green but closer inspection was to show a long low-lying extension of reef concealed beyond it (Pl. 3).

Landing on such an idyllic island as Eilean Tighe was preceded by unexpected argument with John Robertson. Never before had he opposed our wishes but now with stern warnings of tidal dangers he begged us to be on our way. Seeing we were adamant, and speaking almost in a whisper, he then said, 'Yon island is nae place to visit. The wee green folk dinna like it.'

For a moment nobody spoke as we searched his worried old face. Then Wordie very quietly told him that our mission was a peaceful one and that he felt sure no harm would come to us.

After this it was arranged that John should row us to the causeway without any need for him to land and that we should explore and have lunch while John sailed round the islands or fished until he was signalled.

After leaping ashore we climbed a steep grassy bank and deposited our lunch equipment on flat ground about twenty-five yards above the causeway. Our cache comprised a mackintosh to serve as a ground-sheet and a wicker basket containing sandwiches and five bottles of beer. Freed of this load our course continued to the top of the hill where we paused to take stock of the scenery.

One ruin of a tiny shepherd's cottage now overgrown with stinging nettles was the only sign of former habitation but Tom, with his flair for antiquities, was soon excavating a much earlier kitchen midden containing piles of limpet shells and fragments of bronze. Close by was a small pool where grew a few stunted willows, yellow iris and meadowsweet. Buttercups, birds-foot-trefoil, ragged robin, forget-me-not and thrift lent variety to the bright green pasture with, here and there, clumps of marjoram, roseroot, angelica and silverweed.

Whilst I stayed close to Tom in search of insects and spiders the others explored the lower ground towards the reef and when they rejoined us we walked back to have our picnic lunch.

Then came the shock. Lunch had vanished! Disbelievingly, we paced up and down knowing full well it had really vanished without possible explanation. No sign could we find of the mackintosh or the wicker basket with its load of bottles and sandwiches. Even had John dared to land he could not have done so without being seen. Both his alibi and his nature ruled him out completely and there were no other men on the island besides ourselves. Birds, seals and rats could all be eliminated when the nature of the cache and its position were taken into account. Old Harker's tutting disbelief in the impossible was shared by all of us as we gazed around with feelings of awe.

'Let us be away before worse befall,' said John when he had come to fetch us. 'I warned ye,' he added, 'that the wee green folk dinna like intrusion.' He then showed his complete distate for further discussion of an explanation that was so obvious to him.

While John was cooking the lythe which he had caught, a course was set for Skye and I sat on deck writing my natural history notes only to find that another surprise was in store for me. A dozen glass tubes arranged as usual in my waistcoat pockets had all disappeared. On very rare occasions one or two of these have fallen out when stooping, but never all of them and I cannot recall a single instance when I have not noticed the fall and retrieved the tube.

Without insisting that this or another happening some fifteen years later had any connection with the 'wee green folk' a sequel may be worth recording. A friend to whom I had told the story wrote to me when on holiday at Loch Marie saying she had repeated what she could remember to the owner of a steam yacht who was staying in the same hotel. He and his party, it seemed, were extremely keen to visit the 'wee green folk's' island so she asked me for its name. I hesitated. I felt unsure that I wanted inquisitive people to go there, perhaps only for a giggle. However, I ended by sending her the name, but my letter never arrived.

That is the end of the story except to record that nothing happened when Tom, his wife and I returned there with John thirty years later in his boat *Zoraida II.* Mina twisted her ankle on landing but our basket with beer bottles and sandwiches was left intact. Perhaps the wee green folk had moved elsewhere or learned that men bore them no grudge after the islands had been bought by Sir Compton Mackenzie in 1925. True, his tenant's sheep used to vanish mysteriously in the earlier years but this was attributed to raids by fishermen. Strangely enough John showed no anxieties for our safety on this occasion as though he knew that the wee green folk had mellowed or moved.

J. M. Barrie's play, *Mary Rose*, was produced in London for the first time during the following winter. This was about the mysterious happenings on a Scottish island in which Mary Rose vanished and was based on legends associated with the tiny island of Soay Mare which lies in West Loch Tarbert (Argyll). This aroused my interest in Scottish folklore and I searched the literature for similar legends besides making enquiries whenever I revisited the islands.

The Hebrides and quiet spots on the mainland abound with stories of the little people. Those who claim to have seen them say they are seagull-sized or taller, that they are dressed in green, live underground beneath mounds and play sweet music to which they dance at night.

Two boys who stayed out all night on the island of Muck were found next morning in some kind of trance after seeing the little people and hearing their music.

Finlay, whilst in a similar state on Mingulay, fell in love with one of these little folk and was seen by his fiancée being dragged into a bottomless well.

Taransay, off the west coast of Harris, was said by my friend Stephen Courtauld to have the reputation of resenting visitors more strongly than any other island.

A story from Eriskay is of a young girl with a lovely voice who liked to imitate the songs of birds. One night she heard the little people singing beneath a mound so she knelt down to listen until she had learned it by heart. Returning each night she learned new songs which she taught to the islanders and some of these still survive like the Eriskay Love Lilt and the Eriskay Cradle Song. Presently the little people discovered her and she disappeared, but some say she returns when the larks sing.

One unsolved mystery of the past century bears a resemblance to that of the *Marie Celeste*. It surrounds the disappearance of the three lighthousemen on the Flannan Islands to the west of Lewis. There is nothing to be said except that they all vanished without trace and without any known reason. Their beds were made, their next meal prepared and the weather was fair.

In *The Charm of Skye* Seton Gordon has much to say about Eilean Trodday, some ten miles away from the Shiants. Skye itself is full of stories about the little people but Trodday is the isle of the Trolds and should not lightly be visited! About 200 years ago Dr. Samuel Johnson wrote that 'in Troda within three and thirty years milk was put every Saturday for Greogach, or the Old Man with the Long Beard'. The Gruagach is said to be an innocent supernatural who frisks and gambols, but others have cause to fear him or her when disturbed.

Writing in the 16th century Dean Monro wrote an account of his pastoral visits to the outer islands: 'At the north poynt of Lewis there is a little ile called Pigmies ile, with one little kirk in it of ther own handey work. Within this little kirk the ancients of that country of the Lewis say that the said pigmies has been eirded [buried] there.' This island used to be called Ylen Dunibey, the Island of Little Men, but later it became known as Luchruban from the Irish Luchrupain meaning 'the little race'.

Highlanders do not like talking about such things but patient enquiry about the Shiants when in Tarbert (Harris) several years later did provide some interesting information. First, that Shiant means Fairy or Enchanted Island in Gaelic. And, second, that these islands fall in a group which does not provide a welcome to visitors. Many years ago a shepherd and his wife decided to live on Eilean Tighe and built the little house whose ruins we had seen. One evening, while they sat watching their sheep grazing and their little daughter playing in the lush pasture, they suddenly saw her rushing down the hill slope and plunging into the water as though she were being pushed. They felt paralysed as though in a dream and they were unable to move as they watched her drown. Stupefied, they continued sitting there long afterwards until their dog brought them a cake from the house and put it at their feet. 'He is right,' they said, 'we must think of those who live lest they also die.' And with this they got up and abandoned the island, never to return.

As a zoologist I know that primitive animals are driven out by more advanced forms and tend to find their last retreats at the extremities of continents, on islands or in caves, so, if fairies or supernatural beings exist, I can think of no more suitable haven than these lonely islands. For my part, I remain an agnostic, wanting to believe but doubtful. One thing which is sure is that Tom came near to losing his life on an Irish island in much the same way as the shepherd's daughter on Eilean Tighe.

CHAPTER THREE

Further Enchantment in Ireland

THE delightful people of the Irish countryside seemed too good to be true in 1929 as Tom and I motored from Cork to Kerry in search of islands. They were just like the stories of them and, more remarkable still, exactly as they had been described by Giraldus Cambrensis some 800 years ago.

'Tall and handsome' was his description of their appearance. The memory still lingers of the most beautiful girl I have ever seen. She was coming out of a field of sheep as we drove by. Her long dark hair fell on the shoulders of her black dress and framed her perfect features. Blue eyes danced with health and happiness, and her pink complexion was more delicate than any peach. I am glad we did not stop because there is the possibility that had we done so that fleeting vision of the ideal might no longer be with me.

Giraldus spoke of the Irish music which 'cheers the drooping spirit' and to their ignorance of much which seems obvious to an outsider. The reference to their gay jigs is apt and had we been forced to depend on answers like 'down there' to questions about the destination of a country road we would never have reached Kerry.

'Whatever natural gifts they possess are excellent, in whatever requires industry they are worthless' is another quotation! Much as we came to love their charm of manner and kindly intentions, we had to agree with Giraldus. 'Neither did they know anything about the year, the month or the week' he continued. How reminiscent this sounds of their St. Colman for whom, it was said, a cock told the time for his services, a gnat marked his place in the book and a mouse nibbled his ear to keep him awake. For our part, it was soon found necessary to order boats a day or at least several hours before they were needed.

Giraldus spoke of the 'spirit of the brave' and of their wildness in battle. Here his opinion is vindicated by their amazing proportion of the most successful generals in the British Army from Wellington's campaigns to the present day.

Disagreement can be found with his use of the word 'treachery' which combines their love of freedom, their memories calling on revenge for events that had not always happened and their sheer love of a fight for its own sake, whether it be with friend or foe. Tom and I were just leaving a pub at Fenit for a boat we had ordered elsewhere when two newcomers addressed us:

'Will you drink with us?'

'Sorry, we are just on our way to Candiha and already late,' we said.

'Then you must fight with us, you murdering English.'

We drank, and became close friends instantly.

Finally, Giraldus referred to their poverty and sparse food supply. Emigration still continues and we had an experience on the way back to Cork which touched our hearts. Driving along a dusty straight road we gave a lift to a small fourteen-year-old boy trudging wearily with a heavy bag in his hand. This was the beginning of his great adventure, so we felt glad we had been able to help him at its start. Goodbyes had been said to his parents and friends, for ever perhaps, and he was bearing a ticket to America where an uncle had preceded him.

One survival nearly as ancient as the Irish themselves is the egg-shell of a boat known as the curragh in which we were rowed to several of the islands. This has a light wooden frame covered with up to eight layers of buttered ox hides and it is romantic to realise that in such flimsy craft the intrepid Irish are believed to have reached Iceland, Greenland and even America before the Vikings or Christopher Columbus.

We had been to Valentia, Scarif, Church Island, Degenish, the Magharee islands and the Blaskets, in the last of which we had met the 'King' himself, so styled by a tactful Inland Revenue official whose landings had been repulsed by boulders rolled down the hillside. Our next destination was to be the Skelligs which lie ten miles out into the Atlantic from the south-west corner of Kerry.

Skellig St. Michael looked like a sharp pointed pyramid across

Skelligs

the sea and as this was to be our destination there was no time to examine the massive colony of gannets on Little Skellig.

Having landed at a small quay beside a spacious cave an ascent was made by means of about 600 ancient stone steps, or what remained of them, to some fairly level ground just below the ultimate pinnacle. Cushions of sea pink were ablaze with colour; puffins, razorbills and kittiwakes set up their usual clamour around us.

On the platform itself was a group of unoccupied monastic dwellings thought to have been erected in the 6th century not very long after St. Patrick had established Christianity in these parts. Remains were visible of the original little oratory destroyed by the Vikings in A.D. 823 and of the newer one which had taken its place. Close at hand stood an ancient stone cross and a number of old tombstones marking the graves of monks who had died there. Five bee-hive cells, rounded and constructed with mortarless dry stones were miraculously preserved in perfect condition (Pl. 4).

The small colony of monks who had chosen this beautiful spot in which to live and die in a spirit of self-denial and purification must have been severely tested. Contacts with the mainland would have been infrequent, so they must have depended largely on sea birds and fish. Atlantic gales in the winter time would have tormented them with lashing waves which brought cold spray to the very summit while winds would have whistled through every crevice in their cells not securely covered with turf.

Everything was so peaceful that I climbed down to a concealed lighthouse on the island's outer flank to see if the night could be spent there. Meanwhile Tom pursued his archaeological study of the ruins.

The lighthousemen gave no encouragement. Nobody, they said, was allowed to stay without permission from Trinity House in Dublin. Nor would anybody in his senses want to stay there. The island was haunted and things often thumped on their own bolted door at night while the banshees wailed from the graveyard.

Surely, I suggested, this would be the shearwaters, but they replied with scorn that they knew all the birds as they had nothing else to study. It could be the monks with uneasy consciences, or it could be the poor souls whose lives they had been unable to save after a shipwreck a few years ago—or it could be something else.

Tom was sitting on the ground looking pale and shaken when I returned. Before I could speak he said: 'For God's sake let's get off this island at once.' And then he added: 'Something nearly pushed me over the cliff when I was climbing to some nettles just below the oratory and after I had got back to Christ's Saddle some force struck which threw me flat on my face.'

Coming from so powerful and fearless a man as Tom this was extraordinary and, on top of what the lighthouseman had been saying, extremely sinister. Our descent to the boat was made with backward glances as we thought afresh of the farmer's daughter on Eilean Tighe.

CHAPTER FOUR

The Outer Hebrides

ONE of the pleasures of sailing to the Outer Hebrides in a MacBrayne boat is provided for me by the number of old friends in the form of islands at which one may call on the way. In 1952 a gala performance was provided by Scotland's champion pipers on their way to the Games in South Uist. I felt exalted and intensely sad in turns as each piper played his martial air or his lament amidst islands set in a silver sea.

Playing the pipes is a thirsty occupation and by the time Loch Boisdale was reached the martial tunes had become wilder and the dirges more melancholic.

Tom and Mina had been excavating a Pictish wheelhouse on South Uist and were waiting to greet me on the quayside. Here I introduced them to a delightful old Englishman whom I had met on the boat and the conversation was so unusual as to be worth recording:

Me, turning to my new acquaintance, 'May I introduce to you the archaeologist friends of whom I spoke, Sir Hereward Wake? Tom and Mina Lethbridge.'

Tom (like lightning before Sir Hereward could say anything) 'I am really delighted to meet you, Sir Hereward, because I have lately been trying to find your splendid ancestor in East Anglia.'

Sir Hereward, with a smile, 'I really must hear all about this, but first let me introduce you all to my host and hostess.'

His host and hostess proved to be Mr. and Mrs. Herman A. Andreae, who had bought South Uist in 1944 and the introduction led to an invitation to dinner a few days later.

While Tom and Mina were busy on board *Zoraida II* next morning, I walked up the stony track from the harbour to the huddle of grey stone houses which comprised Boisdale. People smiled or greeted me from their doorways and it was evident that whatever Tom or Mina might have mentioned about me to one or two was now known to all. News travels fast in Boisdale.

When abreast the hotel poised near the hill top above the harbour I was warmly invited inside by Milly Mackenzie who was a particular friend of Mina. She was a fair-complexioned handsome Ulster woman whose husband Finlay Mackenzie had inherited the hotel from his father. Finlay raised his large military frame from a desk at which he had been writing and told me that any friend of Tom and Mina was their friend. In a voice that brooked no refusal he said the occasion must be toasted with a wee dram. And then, as though his wife's presence reminded him of some great sadness, some pledge of fidelity, he told me that the doctor had made him promise not to touch the whisky himself before eleven o'clock, but that I must have a taste of the best 'Polly'.

A steeply tilted bottle caused the tide to rise half-way up the tumbler before I could say anything to stem the flow, but during the next half an hour—I was determined to leave before drinking really started at 11 o'clock—I was delighted to learn what he meant by 'the best Polly'.

On a black night in half a gale the s.s. *Politician* had been wrecked on 5th February 1941 with 20,000 cases of whisky on board intended for America. I was now drinking some of the strong liquor rescued from the *Politician* as she lay on the reef between South Uist and Eriskay. As I had enjoyed Compton Mackenzie's book and seen the amusing film of *Whisky Galore* which related to the adventures that followed the arrival of this treasure, the drink took on a fresh relish, but I still managed to escape from the hospitable Finlay Mackenzie before eleven o'clock.

Beyond the village I came to moorland where heather skirted numerous lochs and pools. Here breed the small Hebridean grouse and here is to be found some of the best sea trout fishing in Scotland. My greatest delight was to make the acquaintance of the graceful Water Lobelia whose slim pinkish green stems rose from the shallow pools to bear their pretty light mauve blooms.

This first brief reconnaissance produced an echo of what Dr. Johnson had said of the Hebrides in 1773: 'Your country consists of two things: stone and water. There is indeed a little earth above the stone in some places, but a very little.' He was such a master at condensing essentials that I pondered on what he had said and had to agree with what he meant. The view from the harbour was of smooth hard gneissic rocks with heather and other plants striving for life in the soil pockets collected in its crevices. On the plateau above there had been nothing in view except patches of water and a covering of peaty soil over the island's rocky skeleton.

In this climate a plentiful rainfall and coolness throughout the year prevent neutral decomposition of vegetable matter with the result that layers of peat form from the accumulated residuum of acid humus. The struggle to make sour soils fit for cultivation wherever they are deep enough is ever the object of the crofters.

I took the problem which was now bothering me to Tom. How was it, I asked, that invading tribes had fought so bitterly in the past for possession of such unfertile soil? That, he replied, is not really known and still less, the actual people who had surged in at different periods or their precise dates. The islands were certainly populated in Neolithic times and later, from about 2000 B.C., by Bronze Age Beaker men. An uncomfortable period set in about 700 to 500 B.C. when their grazing lands were wrecked by wetness which swamped them with a blanket of peat. A new race or races arrived before the time of Christ's birth who settled along the western strips where grass grew in the sandy soil. These people were of mixed races often called Picts and often referred to as Celts. Much of the lower lying cultivated land was swamped about 300 years later probably due to a rise in the sea level caused by the melting of polar ice. This resulted in some emigration and the survivors were overcome and enslaved by the Vikings when they started settling in the islands around A.D. 800. Norse rule survived for centuries, but since that time the beautiful but destructive sand dunes along their Atlantic coasts have buried more land and many of the early village settlements as they drive inwards.

The day came for our dinner with the Andreaes and with it an unexpected crisis. Tom was in an unsociable mood and Mina was shy so they decided I must go alone laden with excuses and apologies, which I stoutly refused to do. After much argument three rather grumpy friends set out and all greatly enjoyed themselves!

We were received in a large well-furnished drawing-room with windows overlooking fields sloping downwards to the sea. Here the Andreaes expressed their eagerness to learn of Tom's excavations and showed great interest in his descriptions.

Tom started by explaining the difference between the round brochs or blockhouses (built by the Celts perhaps) on South Uist and the later round residential buildings known as wheel-houses of which he had found a cluster buried in the sand just across the island at Kilpheder. The wheel-house he had excavated had a

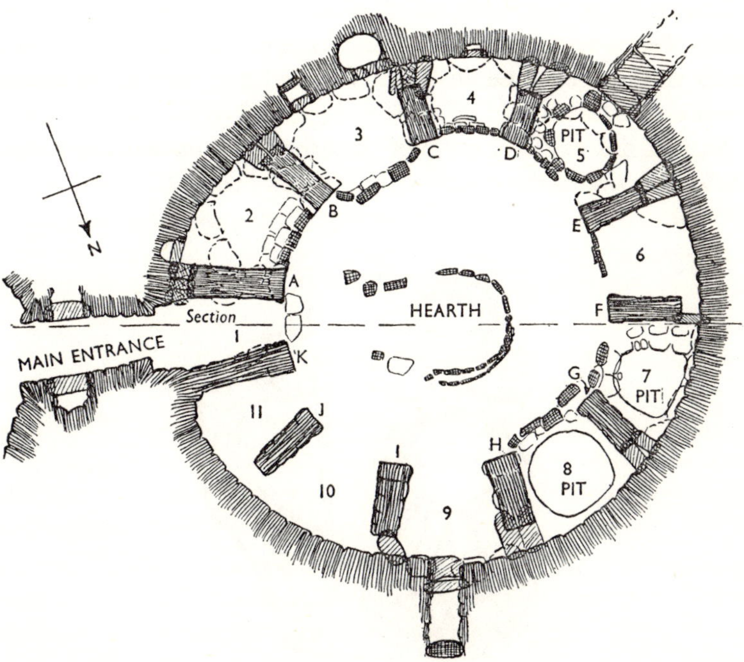

The wheel-house on South Uist

circular room about eighteen feet in diameter with a large semi-circular stone hearth in the centre for communal use. Off this circular room radiated eleven piers of dry stone set in glacial clay to act as mortar. These eleven piers served as walls to a passage of entry and to ten corbelled cells. The piers themselves were bonded with the surrounding outer wall by two heavy lintels about four feet from the ground leaving a narrow passage beneath them from cell to cell. The surrounding stone wall was faced with a thick layer of earth on its outer side.

The way of life of those who lived in the wheel-house could be partly reconstructed from the relics he had unearthed. They used slings, they hunted red deer, seals, fish and birds. They ate winkles, scallops, cockles, limpets and whelks. They kept pigs, spindle-legged sheep and small shorthorn cattle. They had ponies and dogs. Pottery of quite a high standard sometimes bore free-hand drawings of stags besides patterns which suggested an influence from the wares of Roman Britain. Tools, pins and other objects were made of iron and bone. Wool was spun and cloth was woven. Grain was cultivated and ground in rotary querns.

Should this peaceful scene be disturbed by invaders the people could all retire to the stouter defences of a broch.

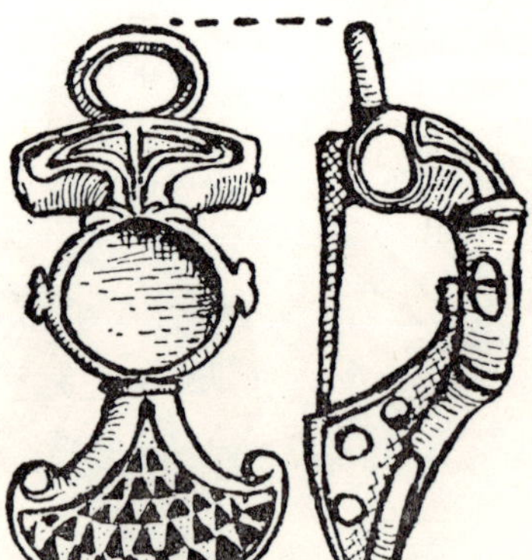

Bronze brooch

Tom's most interesting discovery had been a Romano-British type of enamelled bronze brooch which could be dated to the period A.D. 50 to 200. This was lying on a ledge where it had probably been overlooked when the house had been abandoned and it gave some clue to the date of occupation.

After dinner the Andreaes poured out their tale of woe. Andreae was a London banker whose special claim to fame rested in ocean sailing and in being the first yachtsman in England to use hollow steel masts. He and his wife had fallen in love with the island and with the charm of its inhabitants so they bought it in 1944 with the intention of enjoying the sailing, shooting and fishing whenever they could get there, and of being ideal employers in their kingdom. Now they were disillusioned and nearing despair.

Filled with idealism they appointed a resident factor to develop the farming and to provide the islanders with regular employment and good wages.

Whenever they or their factor tried to introduce new methods or improvements they invariably met with objections. Whenever they sought to employ men on regular employment involving continuity and routine work they failed after a short trial. The islanders whom they had wanted to help had proved to be obstinate, lazy and incapable of seeing what was good for them.

Was it possible, I wondered, for these charming friendly islanders to be so difficult to deal with? Over 300 years ago wise old Sir Thomas Browne had said that obstinacy in a bad cause was just constancy in a good one. In other words obstinacy is not necessarily a bad quality since it depends entirely on a person's point of view. A desire to help and to do good may be praiseworthy and right, but it will be wasted or resisted if it does not fit into the recipients' ways of life and beliefs.

There were many points in common here with the tragic failure we had witnessed for ourselves when visiting Lewis in 1922. After becoming enraptured by that island in 1917, Lord Leverhulme had bought it with the undoubted intention of bringing opportunities, improved health and happiness to the inhabitants. The sincerity of his purpose was disputed by none. He bought drifters and trawlers, erected canning and kippering sheds, built model houses for his employees, made roads and laid out large farms where milk was to be produced in abundance and vegetables grown for the first time.

The island became his in May 1918. His dreams were shattered in less than three years after much money had been spent in pursuit of his ideal. Here, I suspect, the cause of failure was essentially the same as that encountered by the Andreaes in South Uist.

The people were small crofters who overcame their stern environment by turning their hands to several occupations in the course of a year. After ploughing, sowing and peat-cutting during the early months they passed on to fishing, road mending and other tasks before reaping the harvest. Then there was more fishing. Thus the men were not just farmers or fishermen or labourers seeking a regular wage in any one trade. They were all three. And home meant a farm of their own which needed undivided attention at certain seasons.

Lord Leverhulme had not understood the strength of this inbred custom and desire. Nor had he reckoned with the ambitions of soldiers returning from the war in 1918 to own crofts of their own. Loyalty to their chiefs or landlords ranks high amongst the fine qualities of the Highlanders but those in authority are expected to keep their promises and they had been promised crofts by the Secretary for Scotland.

Land tenure in Lewis had never been exactly surveyed or mapped so when land which they had expected to receive was taken by Lord Leverhulme the islanders rebelled and seized it for themselves. Without it the large-scale dairy farming which he had planned for their benefit would become uneconomic. And they did not welcome the regular employment in one occupation that he had thought they needed. Neither the people nor the Government responded to his appeals so his far-sighted plans had to be abandoned. Derelict buildings and harbours already greeted the eye by 1922. Worthy motives are not enough in themselves.

Despite the somewhat grim environment of the Outer Hebrides

they breed strong sentiments and Finlay Mackenzie was one who had felt bound to return there after many years of adventure abroad. After a spell in the Canadian Mounties he had reached the rank of Colonel in the Scots Greys during the 1914–18 War. Now in old age he and Milly managed the hotel where he had been born and there was nobody who came on bad times that did not receive the benefit of their overflowing kindness.

Although Finlay was ready to talk about the famous wreck of the *Politician*, the principal source of first-hand information was a likeable wild character called Angus John Campbell. Angus John had also returned to South Uist after travelling the world in merchant vessels and fortunately for me he regarded Tom almost as an islander himself. Not only had he sometimes worked for Tom on his excavations and enjoyed many a dram with him, but Tom's aunt had been a close friend of a laird long before the Andreaes had arrived. Talks with Angus John combined with the later researches of Arthur Swinson provided a clear outline of the facts of the fairy-tale romance of 'the Polly' so delightfully described by Sir Compton Mackenzie.

The s.s. *Politician* was an 8,000 tonner and her master, Captain Beaconsfield Worthington, was hugging the west coast of Scotland on the way to the northern tip where he was to meet a convoy. The submarine menace was severe in 1941 and the run to America hazardous.

The night of 5th February was intensely black and, some say in a gale, others through compass deviation caused by a curious intrusion of volcanic rock at the south-east corner of South Uist, she plunged off course on to the rocks in Eriskay Sound. Her speed and the waves lifted her securely on to the top of a ridge of rock.

The crew all managed to row ashore and inspection in daylight showed that she could not be refloated. A salvage crew arrived a few days later to rescue what they could of the cargo but that delay was of vital consequence because by that time the islanders had cut a plate out of the side of the hull to ensure quicker loading of their boats before the salvage crew emptied her of precious cargo. They did well for themselves that night but when the tide rose water flowed in to a depth of five feet.

This combined with gales made the salvors' task difficult and they finally retreated with some 500 tons of mails, cars, bicycles, cotton goods and a report that they had saved all they could. The rest of the cargo was largely submerged at high tide by oily water.

Indignation with what followed was voiced by Angus John when he quoted the time-honoured law of the islands that once the salvage men left a wreck, that wreck belonged to the islanders. They were willing to take far greater risks than any salvage men in rescuing things from the sea and when the news spread that beneath the oily water and filth in the holds lay thousands of cases of whisky a fleet of small boats gathered from far and wide despite the high seas and hazardous rocks. They came from Eriskay, South Uist, Barra, North Uist and even further afield to recover the treasure which providence had provided. These were to be the most exciting days of their lives.

The importance of the event may easily be underestimated by anybody other than a Scotsman. Whisky is not only a drink; it is *their* drink distilled from their barley mingled with their peat-flavoured water. It is part of their lives, their traditional pleasure and medicine, almost their life-blood to which they turn in sickness and health, in celebration and adversity. A distant Government in London had introduced fiendishly high taxation and lately deprived them of all but small quantities by exporting the bulk to acquire dollars. The wreck was an act of God and they must risk their lives if necessary to rescue this treasure from the sea.

The wreck moaned and heaved in the swell but stood firm. Reconnaissance showed that they must drag the cases out of the black oily waters in the hold at low tide with hooked spears which were rapidly improvised, and then haul them on deck with rolls of calico from the other cargo.

An enquiry soon emerged in the form of a highly conscientious resident Customs Official from Mull who deemed it to be his duty to prevent the law being broken by the consumption of duty-free whisky. Wind of his visits to the wreck usually got abroad through the need he had to hire a boat, but his frequent house to house search for contraband accompanied by the less eager resident policeman whom he enrolled to assist him had to be guarded against. Directly the cases were landed they were conveyed across the islands to the machar (sandhills or bogs) where they were buried for later use save for the few bottles which were needed to quench immediate thirst. House to house searches brought no reward thanks to the ingenuity of the islanders. Stories were gleefully related of the old lady who filled the chamber pot under her bed just in time to escape discovery of her small store; of the man with a piped water supply who filled his cistern and drew whisky from his tap after the danger was past; of the invalid who filled his hot water bottle; and even of the dead man whose body in a coffin was temporarily replaced by bottles while he was lain under the bed.

In the future some archaeologist like Tom may come across cases of well seasoned whisky in the sandhills or bogs for it was not easy to remember exactly where a cache had been made. Swinson tells the sad story of the man, seriously ill in hospital, who asked after his wee doggie whenever he recovered consciousness. Not until he was well on the way to recovery did the doctor tell him that the dog of which he was so obviously fond had died. This news nearly had disastrous effects on the patient who then blurted out with despair in his voice that the wee doggie alone knew where his whisky was buried.

A story from Angus John was of the priest visiting Eriskay who took as his text 'Thou shalt not covet thy neighbour's goods.' 'Some of you,' he said, 'may have ten cases of whisky. Others may have one case. But I have none.' 'None,' he repeated miserably as he paused to let his eyes wander the length and breadth of the pews where his congregation were seated. His harvest from the sea was plentiful that afternoon.

The joyous treasure hunt continued for weeks and Angus John, without holding the record, accumulated some 300 cases himself, but during April a sourness began to descend on the island. The adventure had gone on too long. The heavier drinkers were beginning to feel the effects of whisky galore. Some, after a generous start, were beginning to make a business of selling bottles for ten shillings each. The normal rhythm of life was affected and the season's work was piling up. The irritation of Eriskay inhabitants at the continued poaching on what they regarded as their wreck was developing. Suspicion that some islanders unknown had given information to the Customs Officer was giving rise to a new feeling of mistrust, especially after several islanders had been caught red-handed and were awaiting trial on charges of theft. Renowned for their honesty, this particular charge caused great indignation. From whom had they stolen anything? The owners, the insurers and the salvors had officially abandoned the cargo so the islanders had only saved it from the sea at the peril of their lives. Their right to do so was sanctified by unchallenged traditional custom yet now they were accused of stealing.

The lonely Customs Officer was obstinately doing what he conceived to be his duty but his arrests provided the Procurator Fiscal in Lochmaddy with a dilemma evident in the verdicts he pronounced. Some were fined a few pounds; a few served up to two months in gaol; many others who had been caught red-handed were found not guilty on this charge of theft.

Although the spree was over, the fairy story continued until the bitter end. After months of excellent work helped by divers a sub-contractor patched and riveted plates over all the holes and reported the ship was afloat and ready to be towed to Rothesay for her steel to be retrieved in the war effort. The contractor, however, was less optimistic and ordered that she be beached on a sandbank north of Calvay near the entrance to Loch Boisdale. The compressed air was released and she sank down not on to sand but on to a concealed rock which broke her back!

And then the final outrage. Before contractors arrived with underwater acetylene cutters to collect what remnants they could for the furnaces at Rothesay, six charges of dynamite were detonated in the hold with the express purpose of destroying the twelve thousand bottles of whisky still remaining there.

'Dynamiting whisky!' Angus John is reported as saying to Swinson. 'You wouldn't think there'd be men in the world so crazy as that.'

CHAPTER FIVE

Return to Scotland

THE western islands are full of tangled history and unreliable legend. Some is romantic and some of bloody battles with pirates or invading forces whose race and origin is often uncertain. There were the Irish, the Celts, the Picts, the Vikings and others unknown. Some were fair and others dark, some were tall and others short. From this mixture came the modern Scot, but despite national pride and clan loyalties we may sometimes spot a throwback with what is believed to be the image of a Celt, a Pict or a Viking.

I like the story told by Arthur Geddes of the French skipper touching at Kintail 300 years ago enquiring of a short dark man to what clan he belonged. 'The Clan Rae,' was his answer, 'but we are of two kinds, the tall fair MacRaes and,' a little pityingly, 'the short dark MacRaes.'

'And what are you?' asked the skipper.

'I,' said the small black Pict, drawing himself up, 'am one of the tall fair MacRaes.'

'God help the short dark MacRaes,' thought the skipper.

Partial segregation of the races and of the clans gave rise, no doubt, to some of the travellers' tales of 'giants' and 'dwarfs' just as they did elsewhere. It needed a Darwin with his tape measure to correct or spoil the tales of a dozen earlier travellers who had reported huge giants in Tierra del Fuego.

The segregation of races in Scotland was not of course complete and one comes across reports that not many centuries ago it was the privilege of the laird to have every girl within his domain immediately before her marriage. No doubt it was some custom of this kind to which Solinus referred in the 3rd century when speaking of the 'Hebudes'. A translation of his writings says that

0 10 20 30 40 50 Miles
Flannan Is.
Callernish
LEWIS
North Minch
Lochinver
Suilven
Summer Is.
Taransay
Tarbert
Shiant Is.
HARRIS
Eilean Trodday
Little Minch
North Uist
Gairloch
Kingsburgh
L. Torridon
Isay
South Rona
Dunvegan
SKYE
South Uist
L. Eynort
Cuillin Hills
Broadford
Kintail
Lochboisdale
L. Coruisk
L. Scavaig
Eriskay
Canna
N
Barra
Rum
Mallaig
Eigg
Mingulay
Muck
Ardnamurchan Pt.
Tobermory
MULL
Oban
OUTER HEBRIDES
Isles of the Sea
Scarba
CRINAN CANAL
JURA
Gourock
West Loch Tarbert

'the king has nothing of his own, but taketh every man's. He is not suffered to have any woman to himself, but whomsoever he chooses he borroweth for a time and so others by turns.'

If Lewis were not so isolated the Standing Stones of Callernish would be far better known than they are. Though not the equals of Stonehenge they must rank high amongst remaining Megalithic monuments in Britain and must have marked the headquarters of some unknown religious cult, probably related to sun worship and revering some great chieftain whose bones were discovered on the site in 1858.

The Stones of Callernish

Nineteen stone monoliths remain to mark an avenue 27 feet in width and 270 feet in length by which one approaches through grass sprinkled with gentians. This avenue runs from North to South and brings one to a circle about 37 feet in diameter formed by thirteen upstanding monoliths. Both to the east and to the west a row of four more extends outwards from the circle, while to the south a few others indicate the former existence of another avenue.

By standing beside the tallest central stone, its height above ground was reckoned to be 16 feet and its total length to be not less than 21 feet allowing for at least 5 feet below ground level. That such gigantic monoliths could have been cleft from the hard Lewisian gneiss before being moved into position and erected here is a feat which gives rise to some wonder.

Tom's flair for discovering archaeological sites has never

impressed me more than in Harris. As schoolboys together at Wellington a Roman site had been found on Caesar's camp after he had noticed a single coin in a rabbit hole. At Mildenhall a slight staining of the sandy soil below the turf marked for him the burial of a Saxon for my spade to excavate. Here in Harris, however, he just pointed to a break in the sandhills below a high crest and said that that was just the kind of spot where Viking remains would be found. Sure enough they were strewn all over the surface of the firm floor from which high winds had shifted the loose sand that normally overlaid it. Nothing really exciting was found like the famous Uig chessmen carved from walrus tusks, but nicely shaped bone winkle pins, iron bosses, pottery fragments and heaps of sea shells at least gave information of former Viking habitation.

Nestling between northern Skye and the mainland is the now unoccupied island of South Rona. A small neat semi-circle of clear water concealed by a rocky islet provided the snuggest little harbour imaginable. A ghost island like this with abandoned church and houses is no happy sight but our purpose was chiefly to examine an isolated cliff cave on the eastern side which Tom had spotted from the sea. After striding across a hill-slope of heather it was eventually found tucked away out of the sight of prying eyes. Only by standing immediately below it could it be seen at all at some 100 feet above sea-level and then, surprisingly, it was quite imposing with a high natural arch of red rock. Six rows of large flat stones were dimly visible inside it, arranged along both sides and capable of seating about fifty people. In the centre and between these was a large level-topped rock which had served as an altar or sacrificial table. Close to the entrance stood a roughly-hollowed font which held water from a natural roof drip.

Old John Robertson, who was a man of near-by Skye, was either reluctant to speak of the cave or else not well versed in its history, but he told us that it was known as 'the Warrior's Cave'. Besides this, he admitted to knowing an old woman, when he was a boy, who used to cross the sea from Skye to attend services there. What services could these possibly be, we wondered, when there were churches on Skye as well as a church just over the small hill on South Rona? Was she a member of some dark cult?

Digging of the cave floor revealed broken pieces of medieval

The 'Warrior's Cave'

pottery with a sprinkling of animal bones which might have been evidence of sacrifices or else the innocent food remains of a holy man.

It is always difficult to unravel the veracity of legends and that of Robert Bruce and the spider is no exception. It was in another cave that fresh resolution came to him as a result of the persistence of a spider in climbing up its silken rope. Unfortunately the Scots are not even agreed as to the identity of this cave, but a visit was paid to one in Arran which is in the short list. A most remarkable thing then happened. Whilst standing just inside the cave I watched in amazement the faltering footsteps of a spider silhouetted against the light as it climbed with some setbacks into the web which I had damaged during my entry. This could have been exactly what had happened to Bruce perhaps in this very cave, and perhaps when watching an ancestor of this very spider!

The spider is called *Zygiella x-notata* and it builds its web across the windows of millions of houses in England so, if it were thought to bring discomforting memories to the English, they can console themselves by reflecting that *Zygiella* now obliges them by catching many a fly or mosquito intent on entering their homes.

When my friend Janey Lindsay had the thrill of inheriting from her godmother, Elspeth Campbell, not only her house near Inverary but also its unknown contents, some inconspicuous cuff links proved to be gifts by Prince Charles Edward to a Campbell for services rendered in 1745. The Campbells were mostly against him so perhaps the exception, Donald Campbell of Glenpean, was the original recipient of these mementoes.

Those desperate months spent by Bonnie Prince Charlie dodging his enemies from island to island have often been described. In retrospect the most remarkable feature of his adventures was that he should have been landed by a French frigate, alone and without troops or attendants, on the tiny island of Eriskay when making his attempt to win a kingdom. And yet, where he failed, one can point to St. Columba's arrival from Ireland in the 6th century on the island of Iona from which he was to succeed in spreading Christianity through southern Scotland and down into England!

The flower which Prince Charles is said to have brought from France and planted on Eriskay is still honoured but, alas, this is one of those romantic legends which it is impossible to accept. The pretty pink convolvulus (*Calystegia soldanella*) may prosper on Eriskay but it is also a common wildflower on neighbouring islands and the mainland.

An authentic link with Prince Charles was provided during Dr. Johnson's famous tour of the Hebrides with Boswell when he met in Skye a 'little woman, of a mild and genteel appearance, mighty soft and well-bred'. This was none other than Flora Macdonald who had disguised Prince Charles as her maid and contrived his escape across the Minch to Skye. According to Boswell, Dr. Johnson not only conversed with her but occupied the bed at Kingsburgh in which the Prince had slept thirty-eight years earlier.

Visits to Dunvegan have always caused me to gaze with interest at the small green island called Isay in the harbour's mouth. MacLeod of Dunvegan Castle offered to give it to Dr. Johnson, whose amazing range of conversation had impressed him so greatly, provided he would promise to live there at least one month in every year. Like many another man, Dr. Johnson enjoyed the dream of an island existence and the dream, while it lasted, softened him and made him a schoolboy again. Boswell was amazed. He had never seen him in so frivolous and gay a mood and he reported Johnson as saying 'how he would build a house, how he would fortify it, how he would have cannon, how he would plant, how he would sally out and take the Isle of Muck; and then he laughed with a glee that was astonishing and could hardly leave off'. The idea of being called 'Muck' in Scottish fashion after his conquest of that island caused him further delight.

How Dr. Johnson can ever have brought himself to leave London for the rigours of travel and the discomforts of the Hebrides is a matter for great surprise, but it had been brought about by the influence of Boswell and by the reading of a book by Martin Martin published in about 1695.

Finding myself greatly intrigued by one statement in Martin Martin's book, I decided to start an investigation. He reported that the inhabitants of Mull chewed the root of the charmel to prevent drunkenness in its damp climate where much whisky was drunk. Modern botany books made no reference to any plant with this name or its spelling variants, so old ones were consulted and presently another interesting claim was unearthed. Anne Pratt said the chewing of the dried root gave an extra relish to the whisky! Further probing led to its identification as a tuberous pea with a pink flower known as *Lathyrus montanus* which is also common in England, and the swollen nutty roots were found to have a not unpleasant flavour rather like a radish!

I mentioned these enquiries during a radio broadcast in 1958, suggesting lightheartedly that a plant which enhanced the flavour of whisky and prevented drunkenness deserved special investigation! A few days later a parcel of two bottles of whisky arrived from the Distillers' Company saying 'while our Company

Suggestion Scheme does not permit of awards being made to non-employees, I am sending you a small sample of raw material on which you can undertake some direct experimental work, with and without the addition of chamaille'.

*Charmel (Lathyrus montanus)**

In a letter of thanks I suggested that fields with light rich soil should be planted with the tubers and harvested for the manufacture of cocktail biscuits. 'Diplomat' pills could also be swallowed before attending dinners where much drinking was expected. The idea still awaits exploitation!

Not far south of Dunvegan lies Loch Eynort from the entrance to which there is a glorious view of the Cuillin Hills piling up beyond the glen at its head. Tom and I decided to walk along the coast from here to Loch Scavaig which would be a good starting-point for a climb on the following day.

* A part, greatly reduced, of a complete drawing of the plant from *Drawings of British Plants* by S. Ross-Craig.

We had passed a ruined dun on the headland and were continuing along a turf-covered raised beach when Tom suddenly pointed to an incongruous clump of nettles growing below the old sea cliff some distance above us. 'That,' said Tom, 'must mark an old kitchen midden.' And then he added, 'Stinging nettles always thrive on rubbish heaps, even prehistoric ones,' which made his deductions as simple as those of Sherlock Holmes once they had been explained.

The nettles grew just outside a small cave which had not been visible from lower down the slope and his diagnosis was confirmed when a layer of limpet and winkle shells had been uncovered. Flat boulders on either side of the cave represented seats and a narrow trench disclosed four layers of soil. The bottom layer was dark and undisturbed. The third yielded small flakes of chert and jasper imported from the mainland which gave evidence of neolithic habitation. The second was of ashy soil which bore signs of a later occupation with fragments of pottery and iron. The surface layer was barren.

Loch Coruisk was soon reached from the anchorage in Loch Scavaig, but on this particular day its famed beauty left a feeling of melancholy produced by the deep shadows on its still waters from the crumbling grey mountain slopes that hemmed it in.

A bare track beside rocks of fine-grained gabbro led to a corrie abruptly terminated by a vertical razor-edged wall of rock which we skirted before finally reaching the summit of Dubh-nada-dheinn.

On clear days a view from the summits of the Cuillins is as memorable as that of their own blue beauty from a distance. At close quarters their bare peaks look sharp, cruel and magnificent. Out to sea is a pattern of islands and to the east the mainland's indented coastline with a jumble of high mountains filling the interior.

Canna provided a great contrast to Skye. It was small, low-lying and insignificant, yet its modest homely attractions would hold me longer than any other Scottish island. Its dwarfed heather-covered ridge of hill was high enough for walks and most of the lowland was of that strangely vivid green herbage also met with on the Shiants. Flowers of great variety grew in the pastures including orchids and the beautiful white Grass of Parnassus alongside the yellow globe flower. The island was deeply indented by the sea and the rock pools were a delight with many of the sea shells and stones encrusted with red coralline growths (*Lithothamnion calcareum*). A little skill with a hook provided a lobster for lunch.

This is a homely island and during a visit in the early 1950's we got to know and appreciate the hospitality of the John Campbells in the main house. Besides tending his sheep he was a great scholar with a reputation for his knowledge of ancient Gaelic and his study of the island's butterflies. More surprisingly, he and his wife kept unusual breeds of cat and he wrote books about them. They had met some thirty years earlier when she had come over from America as a student to study Hebridean music and she had remained there happily ever afterwards. Her life's work was just being published.

One of the sad things about the islands was the way in which the young people went to school on the mainland and stayed there to take jobs, leaving the older generations at home to be visited only when holidays permitted. Progress causes decay and where music was concerned Mrs. Campbell had saved the words and tunes of many a Hebridean song just in time by recording the old peoples' voices. We listened to several of these on her gramophone and got a great surprise. No affinity to traditional Scottish songs was noticeable, but their wild music seemed to resemble much more closely the country music of the Eastern Mediterranean.

Their archaic melodies to accompany work tasks had been passed on from generation to generation, perhaps for thousands of years since the arrival of the Picts. The words were often more modern to bring in the historic events of later centuries, except for the choruses which were often unintelligible sounds.

Many of the songs were 'waulking' songs, songs sung by teams of women during the long 'waulking' or fulling process of the handmade tweed as it lay stretched on wooden planks.

One of the oldest and most attractive from Barra, whose title could be translated as '*One Day as the Knight was Drinking*' told of a knight who pretended to be dead in order to test his wife's love and fidelity. Happily she vindicated his hopes with her grief and by summoning the greatest in the land to his funeral. Mrs. Campbell provided me with a translation:

> One day, as the Knight was drinking
> Ho ro, hug a hug O,
> In the inn, he and his wife,
> Hug a bhi a seinn dug ho ro,
> The Knight cut his finger
> Until he reached the grey bone;
> The blood ran out in a stream to the ground,
> His breath departed like a mist.

Believing him to be dead she went on to send for his friends and to extol his greatness:

> Send for all his friends
> Send for his kin with speed
> To take his body to the graveyard,
> To bury him quickly in the ground.
> It was I [she said of him] who had the ship
> That best left the wave behind her;
> It was I who had the dog
> That best put his eye to the hunt;
> It was I who had the horse
> That was best for riding with bridle;
> It was I who had the wife
> Who was most faithful to her husband.

To complete this ancient ballad, converted later perhaps into a waulking song, references were made to the legendary Fionn, to his henchman Conan, and to his dog Bran:

> One day Fionn went to the hunting hill—
> A headless body met him—
> Conan and Conn arose,
> Bran bristled with rage;
> They deafened my head.

CHAPTER SIX

The Orkneys and Shetlands

ONCE upon a time, though we can be more exact and say during the second half of the 9th century, Harald Fairhair succeeded in making himself King of all Norway by defeating the minor chieftains with the help of Earl (Jarl) Rognvald the Mighty. He then set himself the task of welding the petty states into one nation and suppressing pirates and plunderers with an iron hand.

His relentless suppression within his realm gave great impetus to viking raids and emigrations overseas. These had already started as a result of over-population and an innate spirit of adventure, but the movement was now to be intensified and swelled by discontented members of the aristocracy.

The Nordreys, as they called the Orkneys and Shetlands, had already been occupied by colonising vikings earlier in the century and Harald now conferred the jarldom (earldom) on Rognvald as a reward for his services. Not wishing to live there himself, Rognvald installed his brother Sigurd in his place.

Sigurd proved to be a great warrior and soon extended his domain to include Caithness and Sutherland. His end came from the strange mischance of scratching his leg with the tooth of a decapitated Scottish chief whose head had been hung at his saddle. Blood poisoning was the result.

Sigurd's son Guthorm and Rognvald's son Hallad succeeded him as Earls of the Nordreys, but it was not until Turf Einar, another of Rognvald's sons, came into the title that the earldom regained its earlier prestige. He repelled encroachments by other vikings and gained the appellation of Turf through teaching the islanders to cut turf for fuel.

When Harald's sons grew up several of them became the angry

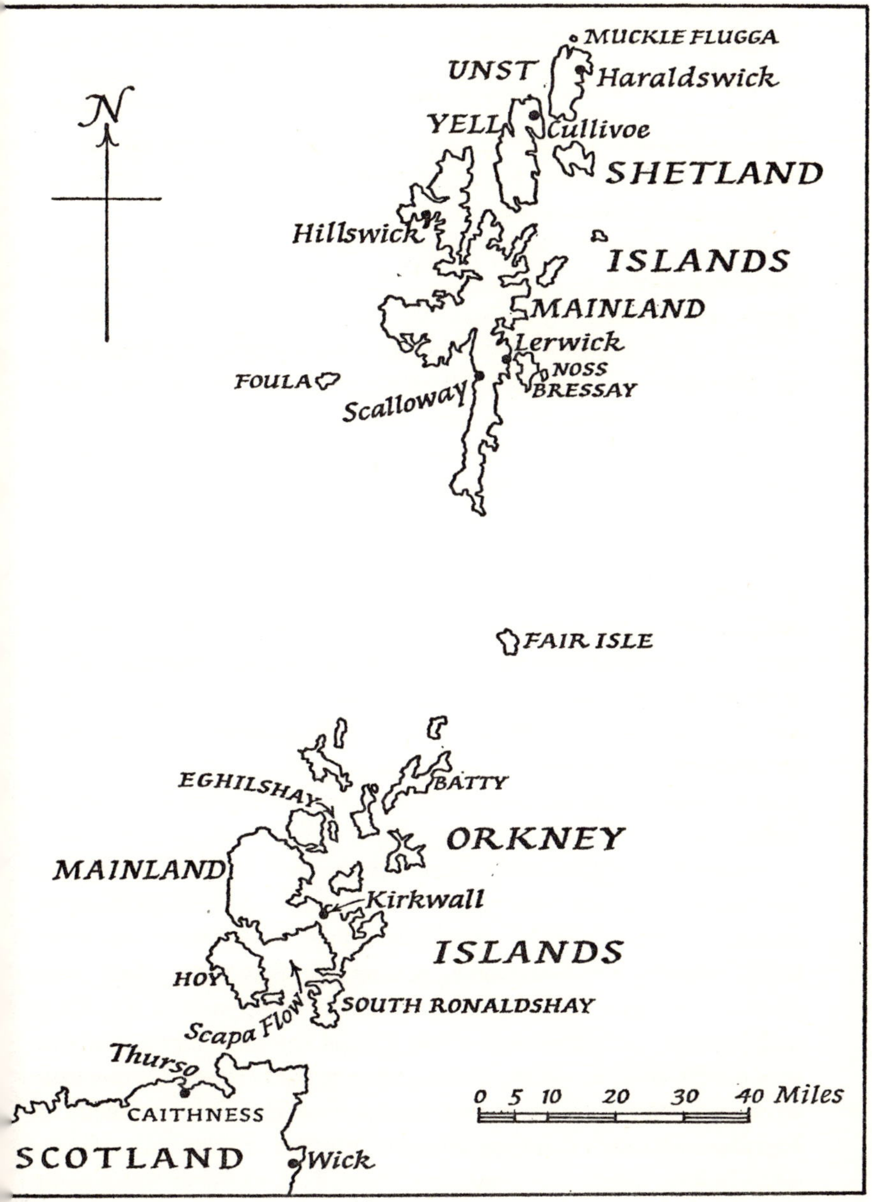
N
MUCKLE FLUGGA
UNST
Haraldswick
YELL
Cullivoe
SHETLAND
Hillswick
ISLANDS
MAINLAND
Lerwick
NOSS
BRESSAY
FOULA
Scalloway
FAIR ISLE
EGHILSHAY
BATTY
ORKNEY
MAINLAND
Kirkwall
ISLANDS
HOY
SOUTH RONALDSHAY
Scapa Flow
Thurso
CAITHNESS
SCOTLAND
Wick
0 5 10 20 30 40 Miles

young men of the period on account of their father's strict discipline. Two of these sons called Halvdan Highleg and Gudrod Gleam set upon the ageing Rognvald and in typical viking fashion burned him and his men to death in their homes. Leaving Gudrod Gleam to possess himself of Rognvald's possessions in Norway, Halvdan sailed westwards with three longships to seize the earldom of the Nordreys.

Turf Einar was taken by surprise, but he managed to escape to Caithness where he organised a counter attack. The ensuing battle was a victory for Einar, and Halvdan was put to death.

This act of vengeance by Einar brought King Harald himself to the Nordreys but he took his son's massacre of Rognvald into account and forgave Einar after sentencing him to a fine of 60 marks. Einar paid the fine himself in return for the cession of properties to him by the islands' odellers (freeholders).

The ebb and flow of power possessed by Einar's descendants continued for many generations and their energy fills one with amazement. Earl Sigurd the Stout, for instance, fell at the Battle of Clontarf in 1014 on the eve of gaining the Dublin throne and his still more illustrious son, Thorfinn the Mighty, became a rival in power of the Scottish King by gaining nine earldoms in Scotland besides the Isle of Man and estates in Ireland. While ruling his scattered dominions with patience and sympathy, he even found time to visit Denmark, Saxony and Rome before dying in 1064.

There was no great earl of equal power to succeed Thorfinn so for many years the Nordreys were divided between brothers or cousins. This was still the position in 1103 when the islands were ruled jointly by Thorfinn's grandson, Haakon Pallson, and Haakon's cousin, Magnus Erlendsson, as dependants of the King of Norway. Haakon was a true viking in spirit while Magnus was gentle and saintly so a clash became inevitable. A meeting to discuss their differences was arranged on the Orkney island of Egilshay with each bringing an equal number of men in two boats but Haakon decided to reinforce his will by arriving with eight ships of war.

The majority decided that only one earl could rule, so in order to save bloodshed the saintly Magnus offered three choices to Haakon. He, Magnus, would go to Rome, never to return, or he would go into exile in Scotland, or, if thought necessary, he would agree to be maimed and imprisoned.

After further debate the decision was reached that one of the earls must die, so Magnus then told his executioner to stand in front of him 'and hew a mighty stroke on the head, for it was not fitting that high-born lords should be put to death like thieves'.

After this execution in 1115, Haakon showed penitence by going on a pilgrimage to Rome and spent his remaining eight years governing the islands with considerable skill.

Haakon's death was followed by a period of bickering among his sons before a kinsman of Magnus known as Rognvald Kali regained the earldom and built a cathedral in Kirkwall in memory of Magnus who had been canonised as a saint in 1135.

Successive Norse rulers continued to battle at home and abroad until finally the Orkneys and Shetlands were ceded to King James III of Scotland in 1472 in exchange for an unpaid dowry agreed at the time of his marriage to Princess Margaret of Denmark. This purely political bargain did not in the least reflect the inclinations of the almost wholly Norse population who still retain pride in their viking ancestry and delight in all reminders of it.

One annual Norse ceremony in January is Up Helli Aa which marks the Julfeast, from which the word Yule in the Christmas season is derived. Originally it represented the return of the sun god to dispel the gloom of the long winter in northern latitudes. A 'guiser jarl' is elected who appears in chain mail and a raven-winged helmet, accompanied by 300 vikings carrying double-bladed battle axes and lighted torches. The torchlight procession conducts him through the streets to the launching site in a large warship decorated with shields and a dragon figurehead. Here the guiser jarl steps out and the vikings fling their lighted torches into the warship which is quickly transformed into a mass of seething flames. In this manner the sea king is sent to his valhalla of gods.

In wilder days of old, it is said that the ship was launched over the bodies of maidens sacrificed to the sea god, but as time went by it was discovered that he seemed equally content with food and wine, so the custom finally emerged of merely breaking a bottle at the launching of a ship.

One special attraction of a visit to the Orkneys and Shetlands sprang from the family tradition learned during my childhood that we were descended from Rognvald the Mighty through his son Rollo (Rolf) who had conquered Normandy after being outlawed by Harald Fairhair. Massive research had lent support to this tradition when it was found that one of Rollo's descendants called Stephen Fitz Hamon had been known as Stephen de Burstow after settling at Burstow in Surrey in the 12th century. Thereafter descent without a break had been fairly plain sailing as his family, soon known as Bristow(e,) had been loyal to Surrey and Sussex ever since.

The way to the islands selected by Richard Lynex and myself was by water all the way from Westminster Bridge. At Limehouse a launch was exchanged for a ship which connected with another at Aberdeen bound for South Ronaldshay in the Orkneys. This had sounded fun but the North Sea was rough and sullen with the result that we saw nothing until Aberdeen was reached. Rail would have been quicker and more comfortable so a day ashore enabled us to arrange the cancellation of our return passage. We were in for a surprise. At the shipping company's office the statement was made politely but firmly that the Aberdeen Steamship Company never gave refunds on return tickets.

Leaving the battle to be refought in London we turned indignantly into a restaurant for lunch, wondering if the unbelieved jokes about Aberdonian meanness really had some substance after all. We had nearly finished our lunch when the naturalist in the party saw to his horror the unmistakable leg of a cockroach in his dish of prunes. That decided us to be tough with Aberdeen in return for their treatment of us and we told the waitress we would not pay for the lunch. An indignant manager arrived. 'But you had almost finished', he said plaintively. 'Yes,' we replied, 'all except its leg.'

Satisfaction with this minor compensation was short-lived. A copy of the *Glasgow Herald* in the Orkneys told us that one Elizabeth Jane Allison had just claimed £500 damages from the Hamilton Cooperative Baking Society for loss of health occasioned from a foreign substance in her currant scone—a cockroach. So Aberdeen had got the better of us again!

I told this story to Maxwell Knight several years later and he immediately capped it with a better one of an unscrupulous friend who had travelled across America without ever paying his hotel bills. He had chosen his hotels with care, always solid respectable family hotels, and, with the help of a few bed bugs from a collection he had brought with him, he was invariably begged to regard himself as having been their guest in the hope that he would not report the unfortunate occurrence.

An impression of the Orkneys from South Ronaldshay was of green islands with low smooth hills and deep inlets or voes. Scapa Flow lay to the North, like a huge lake hemmed in by islands, and a crossing of its placid waters was made to the landlocked harbour of Kirkwall after one night's pause at St. Margaret's Hope.

Kirkwall had become the capital of the Nordreys in the 12th century but, on the whole, it was quaint rather than beautiful and not richly endowed with ancient relics from the Norse period. Circuitous narrow flagged streets led eventually to the Cathedral but this was disappointingly unlovely and scarcely different from many a large parish church on the mainland. A few pillars and part of the western choir were the only visible reminders of Earl Rognvald's 12th-century memorial edifice to his kinsman, St. Magnus, though it was said that the bones of both were enclosed in two of the pillars.

An event which all were discussing in the Orkneys and Shetlands was the sudden appearance a week or two before our arrival of clouds of beautiful butterflies, the like of which no one had ever seen before. Three days later they had all vanished. Letters had been written to the local newspaper, and old ladies and children had captured them as pets to be fed and cherished. A dead specimen showed that they had been Red Admirals, which are one of the butterflies known to embark on mysterious migrations from time to time.

Careful enquiries did not disclose their departure points in northern Scotland, but it was established that clouds of them had arrived in the Orkneys and on Fair Isle on 27th June, and that they had pursued their course to the Shetlands as far as the northernmost island of Unst which they had reached by 4th July.

Leaving the smooth green hills and inlets of the Orkneys for the wilder moorlands and majestic cliffs of the Shetlands, we had intended to stay a few days midway between them on the tiny speck of an island called Fair Isle. Unfortunately, dense mist allowed the ship to go no closer than the sound of its foghorn. Perhaps it was in this kind of weather that the *Gran Grifan* had been wrecked when attempting to sail round Scotland after the defeat of the Spanish Armada. Her commander, Juan Gomez de Medina, and his men had managed to scramble ashore and, during their months of enforced idleness on this tiny island only two miles in length, they had performed a lasting service which brought fame to Fair Isle. They taught the local women how to knit the Andalusian patterns of their homeland which have never been forgotten.

Here and in the Shetlands the wool is not sheared but collected by a process known as rooin. At a certain season the wool becomes loose on the sheep's back and when plucked at this time it is soft and has lasting qualities. Their hardy little sheep provide black, white and grey wool, with two other delicate shades known locally as shaela and mooret (moor red), from a combination of which the patterns of Fair Isle jumpers can be made.

The approach to Lerwick is singularly attractive. Its grey stone buildings rise out of the sea and are arranged along the curve of a bay sheltered by the island of Bressay. From the sea itself the old town is built on a slope with narrow paved streets meandering upwards past jutting unaligned houses to the new town.

Outside Lerwick two brochs stand on hills as a reminder of times even earlier than the vikings. The high smooth walls of one of them curve slightly inwards and then outwards again at the neck to defy ascent from the outside. The closer broch or fort has been excavated recently and the evidence points to occupation by Celts or Picts from 700 B.C. to about A.D. 600.

Broch at Lerwick

A six-mile journey across bleak moorland leads to the old Norse capital of Scalloway and after climbing the Windy Grind there is a delightful view of deep voes (sea inlets), of Tingwall Valley (Thingvollr) and of a loch, now frequented by anglers, with stepping stones to a holm where the open air parliament (the Aething) was held in Norse times.

Words and names are a constant reminder of the past. The peoples' profiles, their speech and sentiments are alien to anything typical of the Scottish mainland and their attitude was exemplified by a conversation along the following lines with one intelligent islander who said he had never been to Scotland:

'But you are in Scotland.'

'Oh no, this is Shetland,' was his reply.

'What you mean, then, is that you've never been to the mainland.'

'But this is the mainland,' he protested, and sure enough the main island, both here and in the Orkneys, is called Mainland.

Lerwick is an important centre for the herring fleet during the summer months and the fish market was bustling with activity. Steam whistles were blowing urgently to announce the arrival of drifters eager to unload their catch, piles of fish were being sorted, measured, auctioned and cured. Some were being packed into barrels and others conveyed to the kippering sheds.

Rows of hefty weatherbeaten girls were gutting, salting and packing fish on long benches with amazing dexterity while still managing to laugh and talk happily to each other. Elsewhere fish were being sliced, washed and hung up in rows in a kind of huge chimney where wood chips, preferably of oak, were fired to kipper them brown without the need for dyes or other artifice.

Fishermen are notorious for the superstitions which demand adherence but one, which decreed ill fortune should clergymen or women accompany the fleet, had recently been dented by a drifter captain who had braved the presence of a girl and been blessed with a record catch of 300 cran or nearly a quarter of a million fish. The news had spread and the captain of the *Lord Barham* from Lowestoft seemed disappointed that we had no girls to bring with us when he agreed to take us out for the night.

After reaching the herring fields in daylight two miles of net were released with large round floats to keep them about three fathoms below the surface. This was followed by an uncomfortable pause of five hours while the moon rose and the *Lord Barham* tested our stomachs with her constant pitching.

The immense task of hauling in two miles of net in an orderly manner now started. At first there was nothing, then a stray fish or two and suddenly a silver mosaic, a gleaming carpet of fish. At this time of year it is never quite dark in the Shetlands and what is known as the 'simmer dim', reinforced by moonlight, produced a beautiful picture of men hauling quantities of shimmering silver fish into the ship's belly. The sight of all this silver lifted their spirits and they sang wild songs as they hauled.

Our flagging spirits had also been stirred and we arrived at

Lerwick in time for breakfast happier though emptier than at the time of our departure on the previous afternoon.

Bressay, in the harbour mouth, is like a stepping stone to the bird island of Noss, so, after being rowed to Bressay and crossing it, we whistled for a ferry. The resident shepherd of Noss soon arrived to fetch us across the narrow sea channel and to direct us up a heather-clad incline to the famous bird cliffs.

Suddenly we came to the edge of a vertical wall of rock plunging 600 feet into the sea. We lay on the grassy brink and peered over. At certain levels an overhang prevented continuous view but every ledge in sight was occupied by countless birds—guillemots and razorbills on one, kittiwakes on another, gannets in dormitories on others and so on. A wild chorus of protest had sounded at our arrival and the restless jostling which took place on the narrow terraced homes must have endangered many an egg or chick.

At this height the sea's continuous swell was smoothed by distance except where it rose and fell on the black tilted slabs of rock at the cliff's base. After a time I drew away from this mesmeric scene and lay on my back in the heather watching a large heavily-built brown bird bullying even the largest gulls in the sky and forcing them to disgorge their fish which a spectacular stoop enabled it to recover before reaching the sea. This was the bonxie or great skua of the Orkneys and Shetlands. On our way down the heather slope our nearness to their nests brought them whistling past our heads at disconcertingly close range.

The sagas tell us that Iceland was discovered by Gardar Svavarsson after he was driven off-course by wind and storm when on a voyage to northern Scotland, and that Floki Vilgerdsson, from the Shetlands, was the first person to go there with the intention of remaining. However, Floki's cattle all died for want of fodder during winter so it was left to others to colonise what he had christened Iceland on account of his uncomfortable experiences there. The large Loch of Girlsta (Geirhildavata) seen on the way between Lerwick and Hillswick was so named after Floki's daughter Geirhilda who was drowned there.

The wild scenery around Hillswick proved to be the most enchanting of any we saw elsewhere in the Shetlands. Here were the highest hills and the deepest valleys while the jagged coastline and offshore islets gave evidence of the sea god's terrific strength. When walking along the cliffs we suddenly became aware of deep rumblings under our feet and of the sound of surf further inland. The sea had undermined the land through a huge chasm and presently we came to the Holes of Skraada where it booms loudly and forces up water spouts on to the surrounding heath in rough weather. Out to sea the small island of Dore Holm had had a 70-foot arch drilled through its rock mass and the weird group of rocky islets known as the Drongs were reduced to teeth and pillars.

I did not copy the remarkable wording on a tombstone in Hillswick churchyard which conjured up a bitterness which had arisen towards the close of the 19th century, but here lay the body of a devout Christian (James Robertson, I think) whose relations could not resist attributing the blame for his death to the local chemist who had supplied him with nitre in mistake for epsom salts. The memory of the chemist's speedy flight to avoid a lynching still lingered.

The highest 'mountain' in the Shetlands called Ronas (1,475 ft.) stands not far away, so Richard and I were ferried across a sea loch to gain what we hoped would be a lovely view from its summit. Down came the rain in torrents from a thick cloud which had anchored on its higher slopes and the kindly boatman was so disturbed at the risk of our getting lost that he insisted on lending us a whistle.

A cairn marking the summit was reached eventually in cloud which obscured the view completely but, as we stood shivering in our waterlogged clothes, Richard performed one of those warming acts so typical of his thoughtful kindness. From his pocket he produced a cigar which he had brought specially for me although he was a non-smoker himself.

The boatman's anxiety for our safety had kept him at his post throughout our absence and he showed genuine relief when we returned. An offer to tea in his cottage had to be refused because of the plan we had set ourselves to cross to the next island of Yell before nightfall.

Disaster awaited us at Ulsta. The tiny hamlet was choked with people attending a land conference. Every house was visited in pouring rain but none could take us in, so we turned back to the post office for advice. Being the Sabbath the telephone could not be used and there was no car in Ulsta to take us elsewhere. At last, but only by bending his religious scruples, the postmaster's kindness prevailed and he telephoned privately to another postmaster fifteen miles away, asking him to send a car to fetch us. The car arrived in due course and we drove from hamlet to hamlet until at last we found a bed in a farmhouse at Cullivoe. It was 10.30 p.m. by that time and we tumbled gladly into the double bed with engulfing feather mattress provided by the farmer's wife.

The ardours and discomforts of the day enabled Richard to fall immediately into so deep a sleep that not even his violent hiccoughs awoke him. These were accompanied by elbow jerks and kicks which constantly found their mark on me because the deep central hollow of the bed inevitably caused both of us to roll down its slope into the centre. During a second night with our hospitable hosts we did better by sleeping in a head to toe position.

After this we crossed to the northernmost island of Unst and were driven in an old Ford to Norwick. Both in Yell and Unst the bleak moorlands were sprinkled with Shetland ponies which were grazing or being used to cart peat turves. High cliffs and short windswept heather provided a wild beauty of their own.

The rocky islet of Muckle Flugga marks the most northerly point in the British Isles but our own pilgrimage ended at three concentric earth circles from which there was a view of Haraldswick where, according to local tradition, Harald Fairhair had landed when he came to punish my kinsman Turf Einar for killing Harald's son Halvdan Highleg.

CHAPTER SEVEN

Lundy and its Chieftains

LUNDY lies twelve miles off the coast of North Devon. It is essentially an elongated slab of granite, three miles long and half a mile wide, with steep cliffs plunging from its flat top nearly 400 feet into the sea. The tableland is reached by one narrow track curving steeply upwards and passing beside an ancient castle poised threateningly near the summit. Beyond the castle lie farm enclosures and a cluster of plain stone houses where the islanders live, about two dozen of them, beside their tavern and lofty church. Slightly isolated, and nestling amongst the only trees in a small combe to the right, is the pleasant home of Lundy's chieftain. The rest of the long narrow plateau is bare except for a lighthouse at each extremity and three stone walls driving across its width.

A summer impression is chiefly of parched greenery despite the island's impressive list of wildflowers. The botanists' particular treasure is a plant with yellow flowers found nowhere else in the world, but this is scarcely a beauty. It is a cabbage! The Lundy Cabbage (*Rhynchoinapis wrightii*) adds to its fame by harbouring its own two special species of beetle.

During a first visit with my friend Richard Lynex in July 1928 the scenery seemed extremely dull but later we found much to admire and interest us. The bracken-covered slopes of the eastern side are attractive but the greatest pleasure was gained along the grander, loftier and more precipitous cliffs on the western flank where awesome turrets and declivities were enlivened by the sound and sight of many birds.

Lundy's commonest nesting birds are guillemots and razorbills followed by puffins, kittiwakes and herring gulls, but there are plenty of others to watch along these cliffs. Hoarse-throated

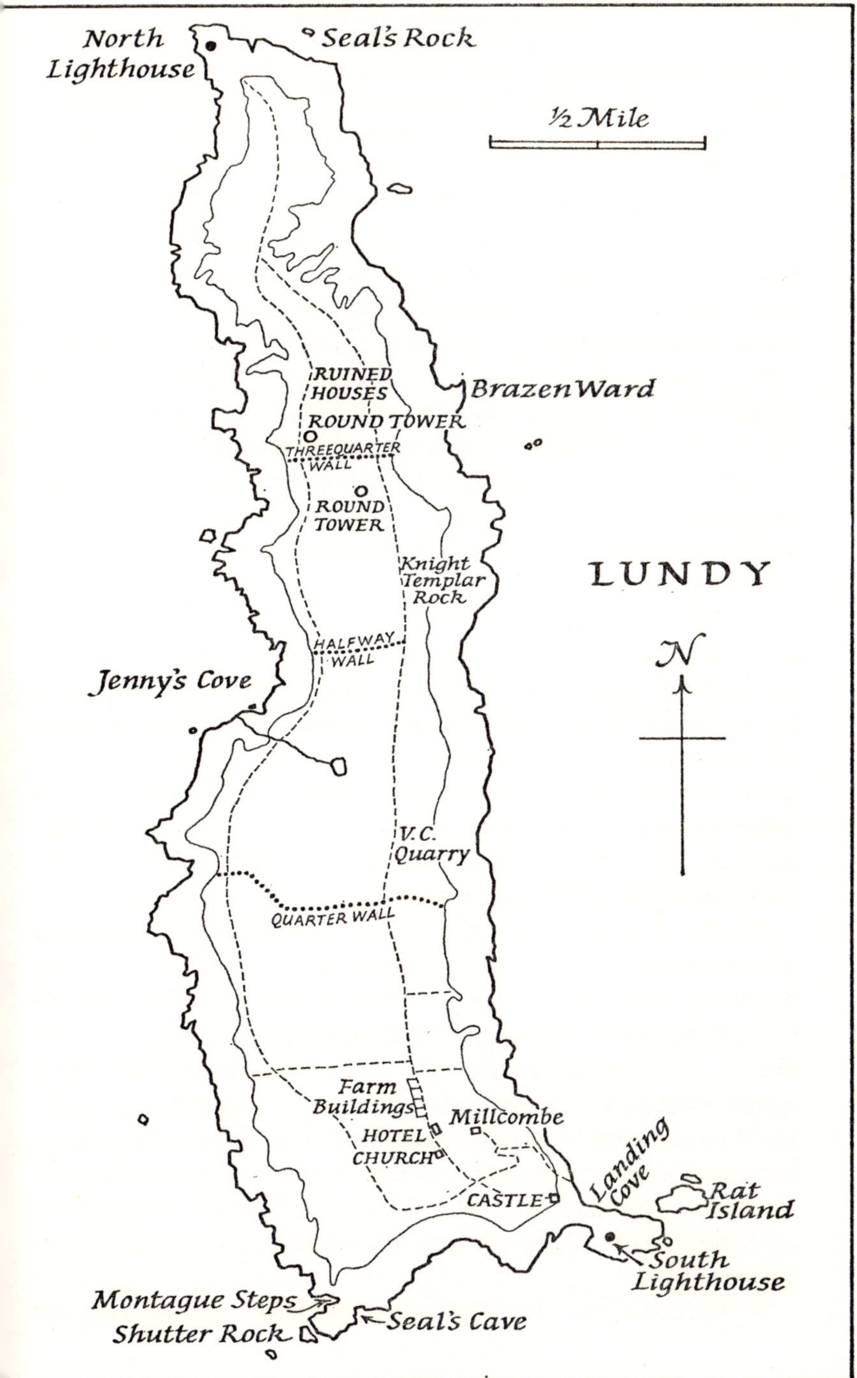
North Lighthouse
Seal's Rock
½ Mile
Brazen Ward
RUINED HOUSES
ROUND TOWER
THREEQUARTER WALL
ROUND TOWER
Knight Templar Rock
LUNDY
N
HALFWAY WALL
Jenny's Cove
V.C. Quarry
QUARTER WALL
Farm Buildings
Millcombe
HOTEL
CHURCH
Landing Cove
Rat Island
CASTLE
South Lighthouse
Montague Steps
Seal's Cave
Shutter Rock

ravens grumble at one's approach, peregrine falcons and buzzards soar overhead and fulmar petrels skim over the sea far below.

During one of our early days Richard and I sat in the hot sun discussing the earliest inhabitants of Lundy. We had just found remnants of these prehistoric men in the form of flint flakes lying on the peaty surface not very far from the northern end. Did they gaze at the same scene of waves pounding Lundy's indestructible granite cliffs, we wondered. And yet, as we began to learn more, it became clear that details of the scene we had regarded as eternal had changed during the intervening millennia. Within neolithic man's gaze no rabbits would have returned his stare. He would have watched and captured gannets from a thriving colony which was completely forsaken by their own whim at the beginning of the present century. Puffins and guillemots were also far more plentiful (and fulmar petrels were not to start nesting until 1944).

An old record tells that no less than 11,000 guillemots, razorbills and puffins were netted in 1787, not for their flesh (which was fed to the pigs) but for the feathers, which brought a handsome income from the mainland, and for their skins from which candle oil was made for their own use. Despite this mighty slaughter, puffins continued to thrive until the arrival of brown rats from some unfortunate wreck in the 19th century. The rats multiplied in the puffin burrows where the eggs and young birds provided them with food in plenty.

Passing to times of recorded history we pictured the outlaws and pirates, for whom Lundy became famous, keenly surveying the waters separating the island from the mainland for ships to repel or plunder. No aggressors could hope to conquer an alert defence force armed, according to the period, with stones, axes, spears, swords, arrows or even guns, as they mounted in single file the only path to the plateau. To this impregnability Lundy had owed its great importance in earlier times and those who held possession were notable for the feelings of kingship which grew up inside them. This has been common to pirate and parson owners alike.

In Roman times the island's strength gave it the name of

Hercules and a relic of their visits was provided by a friendly fisherman of Barnstaple who gave me a small bronze lamp which he had dredged from the sea bottom close to the Lundy shores. Ynys Wair or Gwair which means Island Fortress, was the name given to the island by the Celts. Later the Welsh called it Ely or Hely which is usually thought to commemorate their saint, St. Elen, though to my mind comes the alternative of Helis or Helith who was worshipped at Cerne Abbas, for instance, where their famous hill giant was originally known as Helith. Eventually the name which stuck came from the Vikings in whose Norse tongue Lundy meant 'Puffin Island'.

One of the earliest Vikings to arrive on the scene was Hubba who fought a memorable battle at Appledore in A.D. 878 at which he was slain. It was customary for the Vikings to retrieve their dead so it is likely that Lundy was selected as a convenient burial site and that these burials were excavated in 1850 when farm buildings were extended. The remains comprised a mass of bodies in disarray, a row of seven skeletons and finally two others of obvious importance for whom shaped coffins had been hewn out of solid granite with great labour. Limpet shells, red sherd fragments and pieces of rusted copper or bronze did little to narrow the date of these burials, but some strange round dumb-bell shaped beads are attributed to the 9th century by Dr. D. B. Harden who believes they were of Irish or Hiberno-Danish origin, though others have suggested dates as late as the 14th century.

If the two principal skeletons were Hubba and his son they were certainly formidable warriors. One measured 8 feet 2 inches and the other 7 feet 10 inches! An observer recorded that the shin bones reached half way up the thigh of a farm labourer and that the lower jaws passed easily over his beard and whiskers!

The owner of Lundy at that time had reburied the bones so Richard and I tried to discover them with the only clues available. Searching at various sites just outside the farm walls brought encouragement when limpet shells were unearthed and excitement when a skull was revealed. Alas, we had to be content with two normal-sized skeletons, but interest was derived by finding a

French coin dated 1380 beside the head of one of them. These, no doubt, were the bodies of two French pirates who had been buried here after their last battle.

The adventures of later Vikings are recorded at some length in the Orkneyinga Saga. There is, for instance, the story of Sweyn who 'could not get at Holbodi', a fellow Viking, after he had taken refuge on Lundy. Holbodi had been chased all the way from the Isle of Man by an enraged Sweyn whom he had tried to double-cross there.

When the Viking period gave way to the Norman Conquest of Britain Lundy began to figure in well-documented history. A turbulent Norman family called Marisco gained control and miraculously held it for over 170 years despite monstrous acts of defiance, piracy and treason. One of the Mariscos had been the illegitimate son of Henry I, so perhaps the explanation lies in a combination of the family's influence and the instability of the Crown.

Sir Jordan de Marisco was officially dispossessed of Lundy in favour of the Knights Templars for displeasing Henry II, but the new owners were not allowed to set foot on the island. Defiance of this order was so successful that his son William de Marisco was still in possession forty years later and receiving payment from the King in 1197 for troops he had provided in the war against France. The Knights Templars' rights to the island were confirmed by King John in 1199 but this brought them no nearer occupation than in 1154.

William was a man of spirit so it was probably he who christened a crag, bearing some slight resemblance to a casqued knight, Knights Templar Rock. We can imagine him chuckling in his beard as he regarded this challenging memorial to their impotence.

Although Lundy was his stronghold, William owned ships and pushed his interests far beyond its confines to Scotland and France. Locally he became a notable pirate preying on the merchant ships which used Bristol and near-by ports. Abroad he formed alliances with Scotland and France against the King and it was his contacts with the latter country which probably led to his introduction of rabbits to the island. Rabbits were not endemic to England, and were still a rarity in his time, but by 1284 Lundy was exporting them in thousands to the mainland.

Had he not had a charmed life his alliance with France would have brought it to a close. He was taken prisoner by the English during a sea battle against the French fleet but under the terms of the ensuing peace treaty he was set free by the newly crowned Henry III. Vowing future loyalty to the King he and his family not only returned to Lundy but were allowed to instal a number of stone-throwing machines (mangonels) in 1222 in order to make his defence more sure.

During the next thirteen years William and his son Jordan both died leaving a new and law-abiding William de Marisco in possession of Lundy. Meanwhile kinsmen in Ireland were in trouble. Jordan's uncle, Geoffrey de Marisco, Justiciar of Ireland, had recently had all his estates confiscated for acts of rebellion against Henry III. After being released from prison Geoffrey and his son William went to London to seek the King's pardon and the return of their estates. A clerk called Henry Clement showed such open opposition to their cause that the hot-headed William caused threats to be conveyed to him through his agents. These had no effect and in May 1235 Clement was murdered in his lodgings. William fled and was outlawed but he managed to escape to Lundy where his namesake was living a peaceful existence. His arrival as an outlaw placed this William, son of Jordan de Marisco, in such a difficult position that he promptly left the island and disassociated himself from his fugitive cousin.

William, son of Geoffrey, was now in command of Lundy and black days of terror had returned. Piracy was resumed and attempts to storm the island's defences were all repelled.

Alexander II of Scotland was on the brink of war with Henry III at this time and William de Marisco decided to take sides with Scotland. He sent an armed agent to Woodstock to murder the King and this agent managed to penetrate as far as the royal bedroom before being arrested. William's plot was uncovered and his capture became a matter of utmost importance. This difficult task was given to a Norfolk baron called William Bardof.

Bardof went to Devon and laid his plans with care. Frontal attacks on Lundy were useless so he favoured stealth. His task was aided by alliance with a man detained on Lundy against his will and Bardof now awaited a day on which dense mist shrouded the island. Bardof's small party of armed men then succeeded in climbing the hill unobserved by its defenders and in rushing into the main house where they took William and his confederates by surprise.

The prisoners were carried to London and William de Marisco was sent to the Tower pending his trial in Westminster. In July 1242 he met the horrible fate of being dragged all the way from Westminster to the Tower before being hanged, suspended on a hook and disembowelled.

Death of William de Marisco

Although descendants of the other William continued to have a tenuous interest in Lundy until 1321, their complete control had now come to an end. Henry III sent a resident garrison and in 1243 built a castle which, by some strange twist, is still known as Marisco Castle.

Originally this castle comprised a small square keep with slit windows and an eight-foot gate of entry. Outer walls and a fosse completed its defences. Traces of the outer walls, fosse and the blocked-in doorway can still be seen but no sign remains of the original windows.

Edward II was to seek refuge here from his queen, Isabella, and the rebellious barons in 1326 but 'after beying tossed with contrary winds for a space of a weeke' his ship turned back to South Wales where he was captured and deposed.

Another to seek refuge in this stronghold was Lord Seymour, High Admiral of England, in the mid-16th century. He was accused of conspiracy with pirates and trying to take Lundy that 'he might at all tymes have a sure and saufe refuge'. The penalty was execution.

During the period 1577 to 1701 six generations of the Grenville family held Lundy though not quite continuously. They started with the gallant Sir Richard of whom most of us first heard 'at Flores in the Azores'. The long Grenville rule was a dismal failure because none of them seem to have been residents there for long. The defences were not maintained and the island became a base for English, French, Spanish, Algerian and Turkish pirates. This brought a stern rebuke from Queen Elizabeth who threatened to take the island into her own hands if Sir Barnard Grenville did not do better.

Two notorious English pirates who used Lundy as their base were Captain Thomas Salkeld and Captain Robert Nutt. The feeling of power which Lundy has always exerted on those who possessed it was demonstrated by Salkeld crowning himself Pirate King of Lundy in 1610 and by Nutt promoting himself to the rank of Admiral in 1632. Nutt's dislodgement was swiftly followed by a Spanish raid during which one of the Lundy farmers was killed. The residents had other things to worry about, especially the loss of their sheep, it seems, for 'they burned farms, took away our young sheep and our daughters and left us only old ewes and old women'.

One strange record is of the Grenvilles giving sanctuary to Sir Lewis Stukeley, the man who had arrested Sir Walter Raleigh with whom Sir Richard Grenville had been associated in glory. Sir Lewis found himself surrounded with odium for this deed and he spent his last two years on Lundy after receiving the following reply from James I in 1618 to a letter seeking his protection: 'On my soul', wrote the King, 'if I should hang all that speak ill of thee, all the trees in the country would not suffice.'

In the course of our walks, Richard and I were able to detect at various places on the plateau the traces of round towers which probably marked the sites of ancient watch towers. What is more sure is that Brazen Ward was so named because of brass cannon set there on a platform by Thomas Bushell in the Civil War. Bushell was a delightful character whom Charles I had appointed as guardian when he doubted the loyalty of Sir Bevil Grenville. Bushell had mining interests in Wales and was a genuine island-lover.

Bushell found that the castle was in a poor state of repair on his arrival in 1643 so he rebuilt it at his own expense and later 'defended the Isle against all invasion, it being attempted to have been stormed with twenty saile of ships at one time'. So successful was his defence of Lundy that it had the distinction of being the last loyalist outpost in the kingdom unconquered by Cromwell's men. In the end Lord Fairfax sought to negotiate terms for the island's surrender which Bushell waived until he had had time to consult his King. 'We have perused your Letter,' wrote Charles I in phrases of royal 'wegotism', 'in which We finde thy care to Answer thy trust We at first placed in thee. Now, since the place is unconsiderable in it self, and yet may be of great advantages unto you in respect of your Mines, We do hereby give you leave to use your discretion in it, with this Caution, that you do take example from Ourselves, and be not overcredulous of vain promises which had made Us great only in our sufferings. . . .'

Bushell surrendered Lundy in February 1648, just eleven months before Charles I was executed.

Exactly 100 years later, and some forty-seven after the last Grenville link had been severed, the island was acquired by a colourful rogue called Thomas Benson whose references would have impressed any modern estate agent. Benson had been Sheriff of Devon, a Member of Parliament and now held a position of trust which had given him a Government contract to ship convicts to America.

Benson wasted no time in diverting some of the convicts to Lundy, thereby profiting on his contract and providing himself with all the cheap labour he needed. The convicts were imprisoned in the castle by night and brought out to do his bidding during the day. They are known to have built two of the walls which cross the island and I would suspect they either excavated or at least extended and tidied the mysterious cave still known as Benson's Cave. Benson needed this cave for his subsequent acts and its excavation would have involved such immense labour that only 'slave' labour would have made its construction worth while.

The cave is skilfully concealed from view on the slope just below the castle and has a small entrance, 5 feet by 4 feet 4 inches, which he used to close with a massive door. The cave extends 60 feet into the hillside and is 12 feet high. Benson was now well-equipped with a storehouse to receive contraband which could be smuggled into England on dark nights. Like other owners before and after him the feeling of omnipotence exerted by the island's strength went to his head and he bragged openly of his convict labour and of his secret storehouse to a visiting friend. He tried to impress this friend further by firing a cannon ball at a passing ship which had failed to hoist its flag in salute, remarking that 'the island was his and every vessel that passed it and did not pay him the same compliment as was paid to the King's forts, he would fire on her'.

The visitor reported what he had heard and seen and Benson was fined £5,000 for smuggling in 1753. He took no notice of the fine so the Sheriff was forced to confiscate his estates near Bideford and to send an expedition which seized all the tobacco and other contraband in the cave.

Still undaunted, this remarkable man embarked on fresh roguery in the following year. His ship, the *Nightingale*, set sail for America with a heavily insured cargo of pewter, linen, and salt which was unloaded at Lundy in the night. The *Nightingale* then proceeded westwards for a few leagues and was scuttled.

Fortunately the crew were all saved from their open boats and taken to Clovelly where the Captain swore a statement in support of Benson's insurance claims. The plot was unveiled by a member of the crew and a warrant for Benson's arrest was issued.

Where Benson was lurking is unknown, but he managed to make the long overland journey to Plymouth on horseback and to board a ship bound for Portugal leaving his Captain and other confederates to be hanged.

Benson's life had been so much like a story from fiction that one can hardly believe he twiddled his thumbs in Portugal for the rest of his days. The complete silence which followed his escape in 1754 is in itself so unnatural as to stimulate thoughts of instant shipwreck, of capture by pirates or of his sale into slavery.

Lundy's next dynasty of importance started in 1834 with its purchase by William Hudson Heaven. Heaven longed to live on an island and here he was to build the charming house called Millcombe occupied by all subsequent owners. The narrow track from the landing stage was widened, horses were imported for transport, the castle was reconstructed to house three families, a tenant farmer was found to develop the land, and the slaughter of seabirds was stopped. By 1880 there were 280 acres under cultivation and the livestock included 900 sheep and 70 head of cattle besides pigs and poultry.

Appropriately enough Heaven was a religious man and his son and heir took holy orders. The chapel of Norman or earlier date on Beacon Hill was in ruins so Heaven conducted services in his own home until his son erected a temporary structure which Bishop Bickersteth christened 'corrugated irony' at the time of its consecration in 1884.

The life ambition of the Rev. H. G. Heaven was attained three years later by the completion at his own expense of a Victorian 'Gothic' church with a 65-foot tower. The seas had been stormy for Bishop Bickersteth's crossing on this occasion and he referred to the purgatory through which he had passed on his way to the 'Kingdom of Heaven'.

Again we can detect the feeling of sovereignty descending on the Heavens. They refused to accept the authority of the Customs, or of the Inland Revenue, or of a policeman from the mainland to arrest an islander suspected of murder. They even erected notices denying rights of way to the lighthouses and the post office.

Although the odds against winning some huge lottery are immense, they certainly do not exceed the odds against a particular faux pas over Heaven's church in which I became involved.

Soon after we had gone to live on Ranmore Common, Surrey, an invitation to tea arrived from the squire. He and his wife had just returned from a holiday and they happened to mention that they had seen Lundy from the North Devon coast. Hearing that I knew Lundy they asked for information about it. All went well at first, but suddenly an atmosphere of ice was detected which continued until we left their house. It was only after living in Ranmore a little longer that we realised what had happened. In lively fashion I had said that old Heaven must have been a strange man (I could not remember if I had said 'bats) to spend a goodly slice of his fortune in building a church capable of seating 165 people on an island where the normal population was only two dozen. Unknown to me at the time the population of Ranmore was exactly two dozen and the squire's father had spent a goodly slice of his fortune in building a church only slightly smaller than the one on Lundy! The squire of Ranmore was perhaps the one man in Britain to whom my remarks were vinegar.

One appreciative visitor to Lundy during the early reign of the Heavens was Charles Kingsley in 1849. He was so moved by the awesome beauty of the cliff scenery that he exclaimed in a letter to his wife, 'O that I had been a painter for that day at least'. Tales of the wreck of the *Jenny* with a cargo of ivory and gold dust in what is still called Jenny's Cove may have inspired his account of the *Galleon's* wreck in *Westward Ho!* He chose for his scene the Great Shutter Rock where, strangely enough, the *Montagu*, Flagship of the Home Fleet, was to be wrecked in 1906. Steps cut down the cliff by salvage men are still known as Montagu Steps.

Six years after the Heaven dynasty had closed in 1919, the island was bought by Martin Coles Harman. He was an eminent financier with an empire of companies under his control, a man of the world who came to love the simplicity and beauty of his island domain. Helped by his faithful agent, F. W. Gade, through all the troubles that were in store for him, Harman improved the amenities, introduced experimentally a variety of animals, developed the farm, founded the Lundy Field Society for naturalists, encouraged archaeological research and took over the postal and telegraphic services which the G.P.O. were anxious to abandon.

It must have been the G.P.O.'s action which gave him a series of amusing ideas. In 1929 he introduced a new Puffin currency, with one Puffin equivalent to a penny, and issued both stamps and coins.

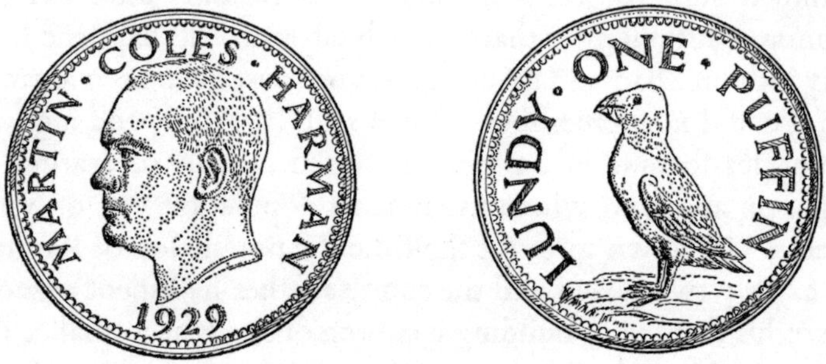

Lundy coinage

The only objection to the puffin stamps, for services rendered in his post office and for carriage between this and the mainland, was overcome by sticking them on the back of the envelope, a practice which still continues.

The issue of coins, however, with their fine head of Martin Coles Harman resembling some Roman emperor raised immediate objection. These, it was alleged, contravened Section 5 of the Coinage Act of 1870.

Harman conducted his own spirited appeal with humour in the High Court and was congratulated by the Lord Chief Justice. He pointed out that no rates, taxes or duties had ever been levied and that there were no Poor Law or Old Age Pension entitlements. Such islanders on the Electoral Roll had been placed there after the informal visit of a Member of Parliament in a visiting cricket team and none had been added since. Acts of Parliament and Statutory Orders relating to Lundy cut no more ice than would an Act of Lundy declaring Devonshire to be part of Lundy, because the island was outside territorial waters. He, Harman, was the sovereign of a small self-governing Dominion loyal to the King, and the Court had no jurisdiction over him.

He lost his case and the fine of £5 was upheld but he had had fun and the satisfaction of knowing that the 100,000 coins he had minted would all be sold profitably as collectors' pieces.

Harman took a warm interest in my researches on his island and he made me a Freeman of Lundy with free board on my subsequent visits, but delicacy prevented my taking advantage of this privilege when next I went there. The collapse and tangled skeins of his financial empire had got him into trouble. Whatever faults he committed my own feelings of affection for him remained.

During the 1939–45 war one of his sons was awarded a posthumous V.C. for gallantry in Burma and an old quarry where he and his brothers and sisters had played as children is still known as V. C. Quarry. Martin Coles Harman was buried on Lundy beside his wife in 1954. Albion Harman, who succeeded him, died in 1968.

CHAPTER EIGHT

Journey to the Åland Islands by Windjammer

IT was of course mad to set out for the Baltic in a ship with no engine or wireless when war was obviously imminent in July 1939 but we decided to take the risk. The opportunity to sail in a windjammer for which my wife and I had been waiting eagerly might never occur again. Surely there was time to fulfil this ambition before the misery of war engulfed us all?

The snow-white *Viking* had discharged her Australian wheat at Cardiff and would now be sailing on her last lap to her home base in the Åland Islands between Sweden and Finland. She displaced 2,400 tons, was four-masted, square-rigged and certainly the most beautiful of all surviving tall ships (Pl. 5).

Our departure from Cardiff was far from the majestic sight promised to a friend standing on a golf course in South Wales who saw us being towed in flat calm all the way to Lundy! Noted for his pungent wit, we wondered what he would be saying to his opponent at the sixth hole.

When fifteen miles separated us from Devon's rocky coast, all thirty four sails were set to make whatever headway was possible in the light breezes bearing us southward towards the Scilly Islands, where we had spent our honeymoon five years earlier.

Save for the Captain and First Officer, the *Viking* was manned by youths in training to get their master's certificate. A trip to Australia and back had made them all highly proficient and their perilous circus act up masts and along spars to reef the sails in rough weather filled us with admiration. In perfect sailing weather she seemed to make about twenty knots and, although she pitched in high seas, it was remarkable how the sails held her firmly against rolling.

The beauty of her sun-lit sails riveted attention as she sped past the Needles and the white cliffs of Dover into the North Sea. There was no engine to vibrate or smell, and only the sounds of the sea and wind in her rigging as we sat on deck in the sunshine. It was just like some lovely dream far removed from the cares of the outside world.

Seven of the passengers besides ourselves were accounted for by the owner's wife, Mrs. Ericson, with her two beautiful daughters, and four Swedes who emerged from their cabins only when the sea was smooth. An eighth was a retired British naval captain called de Vitré who displayed his bachelor's self-reliance by drying his laundry on deck every morning by walking the decks waving a single sock in one hand and a handkerchief in the other. The other sock and another handkerchief would appear on the following day.

It turned out that this gentle old man had been at school with my father and this connection with his past led him to tell me something of his family history. De Vitré's grandfather had been an Admiral in the French navy who had been killed at the Battle of Trafalgar. When the war with France was over the British admiral who had sunk the grandfather's ship had shown great chivalry in adopting his opponent's orphan son. In due course this boy had served in the British navy where his son, our de Vitré, had followed him.

Any amusements on board were simple and unarranged. One of the Ericson girls, for instance, could play a piano accordion and with entire lack of self-consciousness she would often poise herself on a table and entertain us with quiet Scandinavian folk songs. Sometimes this led to a sing-song or even a dance. One night we decided to teach the crew an English game and decided that *Up Jenkins* would be easy. This was a roaring success and we had our reward a few nights later. Framed by a deck skylight we found ourselves entranced by a memorable picture of a game in progress in the forecastle. Hanging lanterns gave shadow and glow to four eager faces on each side of a table with spectators standing in the shadows or leaning forward in excitement to give directions. No sound reached us so the players seemed to be miming the *Up Jenkins* movements. The opponents stared and grimaced while the team in possession of the coin concealed between the forefinger and thumb of one of them went through the motions of 'crab-pots', 'creepy-crawly' and finally 'smashems' as their hands all came down with a silent crash on the table.

War clouds seemed far away as we sailed through the Skagerrak and Kattegat past the narrows at Shakespeare's Elsinore into the Baltic. Then came a shout as two German submarines were seen coming straight towards us. For a moment our hearts stood still as we feared that war had been declared and that the ship would be searched for enemy aliens. Having come almost alongside the *Viking*, they changed course and we breathed again.

By the thirteenth day we were safely anchored at Mariehamn just in time to celebrate the opening of the crefta season. Crefta are small dark red crayfish and we were led into the low-ceilinged cellar of our hotel to join a party given by the Ericsons. A crescendo of noise marked the arrival of huge piles of crefta and healths were drunk in schnapps. Every guest in turn drank our healths to the last drop in his glass and each such personal toast required the same compliment in return. The crefta were excellent but we were glad we had no further than a short flight of stairs to reach our bedroom.

The island was bare, except for one small area of woodland, and chiefly composed of hard smooth rock partly covered by shallow soil. It was far from beautiful and all the houses grouped near the harbour were plain little wooden boxes. Even the cluster of tall ships looked gaunt and lifeless without their sails.

The Ericson home was plainly furnished with spotlessly clean wooden furniture and the absence of decoration or flowers gave it a singularly spartan appearance. After entertaining us in their 'town' house it was surprising to learn that they had a 'country' house built in solitary seclusion on one of the neighbouring islets to which we were invited for the weekend.

This house proved to be a smaller edition of their main home. Summers are short but hot on the Åland Islands so most of the days were spent bathing or sun bathing on the smooth rock slopes beside the sea. Complete relaxation was disturbed by large black snakes, which swam and basked beside us.

Ericson was a short square man with close cropped greying hair and grey blue eyes. He had succeeded in making profits from his fleet by buying his windjammers cheaply from others, by maintaining them at Mariehamn without much cost, by refusing to insure them and by employing trainee crews.

At the time of our visit he was not unnaturally extremely worried by the world situation, but in his talk it was the possibility of Russian aggression and occupation which was disturbing his sleep. The Åland Islands belonged to Finland and not many months had elapsed since an uneasy peace had been declared after some severe fighting.

A week in the Åland Islands was followed by several delightful days in Sweden and Denmark before arriving in England a few weeks before the German invasion of Poland which brought us into the War.

CHAPTER NINE

A Taste of Madeira

AFTER three visits to Madeira I think of it much like H. N. Coleridge in 1825, when he said 'the seasons are the youth, maturity and old age of a never ending still beginning spring'. The weather is usually pleasant all the year round and the island is festooned with flowers at all seasons, blending side by side the beauties of the tropics with those of the temperate regions in gardens large and small.

Those who walk, and even those old or lazy ones who keep to their cars, can enjoy the grand and gorgeous scenery of the central highlands. Besides this there is sea fishing, bathing and much still to be discovered by those with enquiring minds about the history and natural history.

Lest I rave too much about one of my favourite islands, it should be added that it has no classical antiquities or ancient works of art, no sandy beaches and little in the way of night revelry. The Portuguese both in the town and country are friendly and live simple lives, but the island is now approaching a great crisis. Will it emerge from the effects of having direct air flights from Europe spoiled or transformed into something different and more artificial?

English people have cherished Madeira for centuries and played a substantial role in its development, indeed the credit for its original discovery is given to one of them called Robert Machin.

There is much in the history of all nations which is legendary or adorned and not subject to proof, but it is far better to accept at least the outlines of early tales than to create a void by abandoning them. Perhaps the story of Robert Machin has been adorned, and perhaps he was a pirate or ne'er-do-well instead of a romantic hero who eloped with an heiress from Portugal to make the accidental discovery of Madeira, but why should we doubt his existence? It seems so unlikely that the early Portuguese navigators should lessen their own glory by inventing a foreigner who preceded them.

The tale is a sad one. The ship in which Robert Machin eloped with his Anna d'Arfet had been blown off course in a terrible gale when they sighted this unknown island where shelter was sought. They went ashore with a few companions and tarried overlong with the result that the captain abandoned them. Whether he thought they were lost, or was fearful of more storms in these uncharted seas, is unknown, but the shock was too much for Anna and on the sixth day she crumpled and died. With all zest departed, Robert followed her to the grave five days later after expressing a pious hope that some day their joint grave would be found and a chapel erected on the spot.

Their companions eventually managed to construct some kind of boat in which they crossed to Morocco where they were unfortunate enough to be sold into slavery. The years passed and eventually one of them, Juan de Morales, escaped to Portugal.

His tale reached the ears of João Goncalvez, otherwise known as Zargo, for whom an expedition was set up by Prince Henry, the Navigator, to rediscover Machin's island. Zargo was so confident of finding it with the help of Juan de Morales that his ships carried a considerable number of stalwart colonists to be settled there.

On All Saints' Day, 1418, they sighted an island which was promptly christened Porto Santo in honour of the Saints. Having disembarked the settlers he sailed for Portugal with news of their success and returned in the following year with more settlers. This time he sailed on to Madeira, which had only been seen in the distance from Porto Santo during the first voyage, and managed to find the graves of Robert Machin and Anna. Dutifully he built a small chapel on the spot and those who settled there called the village Machico in Machin's memory.

This story was told in 1610 by Francisco Manoel de Mello who claimed to have copied it from an earlier manuscript written by one of Zargo's companions called Francisco Alcaforado.

It seems highly remarkable that such fertile islands had burgeoned in solitude and without inhabitants up to the time of Zargo's arrival in 1418. It must be supposed that the Carthaginians had at least seen them on their way to the Canary Islands more than a thousand years earlier. However, all that can be said with certainty is that two apparently genuine maps, a Medici of 1351 and a Catalan of 1375, give indications of three islands in the Madeira group before Zargo's rediscovery of them.

Zargo and his companions now set about their tasks in Madeira with tremendous vigour. Finding themselves hemmed in on all sides by dense forests and precipitous gorges which made cultivation and lateral communication impossible, they spent years clearing the ground with forest fires. Stone was quarried for roads and terraces up the steep hill slopes, and an elaborate system of aqueducts (levadas) was built to bring water to their crops. We can still marvel today at what they and their successors accomplished in the way of hewn stone surfaces to their roads (sometimes arranged in patterns like mosaics), the amazing tiers of terraces for cultivation up the steep hill slopes, and the hundreds of miles of levadas to bring the mountain rain water to their crops.

The early settlers were helped by slaves imported from Africa and Fructuoso estimated the number at 2,700 in 1552. Slavery, which did so much to develop the island, continued until the close of the 18th century though it was officially abolished in 1775. Nial Cossart's family scrapbook enabled me to trace the history of one of their slaves called Jogo. Bought for 110,000 escudos in 1771, he was valued in the firm's books at 200,000 escudos, in 1786. By 1790 age and infirmity had caused him to be written down to 150,000 escudos and in 1791 he was written off completely as an asset. The comment was made that 'he is no longer to be regarded as a slave' and there is little doubt that his owners looked after Jogo for the rest of his life.

'Slave' is an ugly word which we all detest and which I certainly have no intention whatever of defending, but in practice slavery to a good master was certainly not a living hell and even had its compensations. There was no risk of unemployment!

And in return for their work slaves were provided with food, clothes, medical care and accommodation better than they had been accustomed to in their homeland.

When slavery was finally abolished in Madeira, most of the able-bodied slaves were shipped elsewhere like one called Fortune who was sold by Newton and Gordon to a customer in Jamaica for 100,000 escudos.

The early colonists had imported vines from Cyprus and Crete and their wine soon gained fame. It is mentioned in Shakespeare's *Henry IV*. Charles II encouraged its export to America and the West Indies by allowing direct shipments, instead of via English ships from England like all other goods shipped to the colonies at that time.

The early appreciation of Madeira wine by the British caused many to enter the trade and it was 'convenient' for Alexander and James Gordon to take up residence after the parts they had played in the Scottish troubles of 1745. Here they prospered alongside Francis Newton, the Leacocks and others who came later, including the Cossart and Blandy families.

Madeira became the home of these families and throughout centuries some have retained their British nationality and have sent every generation of children to England for education. The journey used to be long and perilous in the days of sailing ships so we can imagine the Leacocks' anxiety for their children's safety when they prayed that adverse winds would not carry them across the Atlantic to 'those Americans'. This was soon after the American War of Independence!

Even in 1853 the unpublished diary of a young Englishwoman married to José de Franca describes a journey from Shadwell to Madeira in the *Eclipse* which took twenty-three days despite the absence of any storms. Her one complaint was the absence of a cow on board to provide fresh milk for her tea!

Madeira wine has been classified into three main types—sercial, boal and malmsey—but customers often used to specify their exact requirements. I see from the Cossart scrapbook that Captain Rattray ordered a pipe of wine to the following specification in 1787 which required the addition of goat's blood:

> Draw from the pipe 2 gallons of wine in a quart of which dissolve four ounces of isinglas, take a pint and a half (not exceeding a quart) of kid's blood and, as it runs from the animal into a large pan, pour in the other 7 quarts of wine at this time stirring with a whisk, then add the quart the isinglas was dissolved in and let them be well incorporated together with the whisk before it is poured into the pipe—when done—let the whole be mixed for fifteen or twenty minutes by a stick put in at the bung hole.

Part of Australia's early interest in Madeira wine seems to have derived from a belief in its good effects on invalids. A shipment made by Klemm & Co. to the Medical Society of Victoria in Melbourne, brought praise from Dr. James E. Neild in 1876. It was 'suitable for invalids and convalescents', according to his wine committee, and they preferred 'Choice Reserve' as this was deemed 'expressly agreeable to the palate and was apparently borne well by the stomach'. I like to picture these doctors solemnly testing their stomachs!

A catalogue of famous visitors from Christopher Columbus and Captain Cook to Winston Churchill will not be made, but Madeira's popularity as a holiday and health resort owed much to Queen Adelaide when she went to Funchal in 1847 after William IV's death. The fashion which this started provided an enterprising Scottish baker called William Reid with an opportunity to build or control an empire of hotels. Most of these have since been replaced but Reid's Hotel, which he built as the 'New Hotel', still thrives. He started in 1850 with the Royal Edinburgh which was situated near the Public Gardens. The New Hotel came next and later he was to buy the German Hotel on Rua dos Hortos and the Hotel Miles Carmo in Eastern Funchal. When he had gained proprietorship of the Hotel Santa Clara in the hills outside the town, he had far outstripped the Jones's who ran Jones' Hotel in Bella Vista!

Popularity spread to the village of Monte which overlooks Funchal from nearly 2,000 feet, and several fine residences were built there. Monte then became connected with Funchal by a cog railway which survived until the 1939–45 war. Nowadays the ghosts of departed grandeur sadly surround one when sipping the delicious local wine on the shady cobbled terrace overlooking Funchal. A supply of crystal clear water from a shrine-like tap and stone basin built by a former British Consul, Charles Murray, in 1778 is still in constant use and the 15th-century church stands on the hill just above the terrace.

One of the wayside scenes in every village is of groups of women doing their beautiful embroidery work. It is said that 40,000 women and girls are so employed every day. To anyone interested in beginnings it is nice to know that this industry was started by one kindly English woman, Miss Phelps, who thought it would be useful for the womenfolk to have an occupation outside their domestic chores or their work in the fields. When the moon is full the sewing is said to continue into the night and by the time they scatter to their homes all local gossip has been exchanged.

Englishmen, of course, have always arranged their own variety of male gossip centres. Wherever ten or more Englishmen find themselves gathered together they form an English Club but where they met before one was built in Funchal in 1832 is not recorded. Here, also, during the 19th century they formed a British Country Club, a Golf Club, a Floral and Agricultural Society and a Casino. Music, dancing and lectures were also organized but it cannot be claimed that Madeira ever became a really gay place because its visitors were predominantly middle-aged or invalids.

One horror from which English and other Protestant residents used to suffer was that of dying in Madeira. The law allowed only Catholics to be buried on the island and the Protestants had a distinct aversion to being unloaded into the sea off Brazen Head. The Cossart scrapbook reveals that a dying partner in the firm of Alexander and James Gordon made his partners promise to bury him under the floorboards of his office, just below the desk at which he had sat for so many years. A coffin filled with stones was offered to the sea and it was not until many years later when the office was taken over by the Cooperativa that the macabre discovery of his skeleton was made.

Permission for a Protestant burial ground was finally negotiated by Consul-General Nash in 1770 and Mrs. Shipcote, a tavern keeper's wife, was the first person to have the ground opened for her reception.

The first stone for an English Church was laid by Consul-General Veitch in 1814 and it was completed a few years later after much bickering about its architecture.

Apart from riding or walking, two early methods of transport were devised suitable to the steep stone-paved roads. A couple of men could be hired by the hour or day to carry a passenger long distances in hammocks suspended from a pole. The alternative was to be hauled in sledges along the pebbled roads by two men with ropes. In 1847 an Englishman, Major Bulkeley, introduced bullock-drawn sledges. These, like the man-handled ones, survive for the country people in the hills, chiefly for the transport of heavy loads, and also for tourists in Funchal.

As a birthday jaunt my wife and I came down all the way from Monte on a sledge and thoroughly enjoyed it. We had lively steersmen and on the steepest stretches reached at least twenty miles an hour as they perched themselves on the back of the sledge. The polished pebble slides passed between high stone and plaster walls into which large wooden doorways were let at short intervals. There was an Alice in Wonderland quality about gliding past these closed doors with an occasional glimpse of sunlit gardens through any which happened to be open.

Despite the splendid climate a visitor will hear of the occasional summer *leste* which is a hot dry wind from the African desert. One of these started on 17th August 1964 and lasted three days. A haze of fine dust blotted out distant views and a strong easterly wind blew unceasingly. Doors and windows rattled while the temperature indoors rose to between 80° and 86° F. throughout the days and nights. At first it seemed nothing to fuss about, but presently it was noticed that people were gasping, getting nervy or even collapsing. One old Scotsman had two heart attacks before making an emergency departure in a ship bound for home.

Humans were not alone in their suffering. Ants came streaming into bedrooms searching for water in the wash basins. Midges and other flying insects swarmed into the houses.

Unusual moths had been blown across the sea all the way from Africa. Plants wilted and the trees looked particularly forlorn with leaves hanging limply downwards. Quite suddenly the wind dropped and it was all over.

Hurricanes come to Madeira very seldom. The worst on record were those in 1802 and 1842. The latter started with intense rains on 13th October which continued with little respite until the climax was reached between the 24th and 26th. Hurricane winds and waterspouts tore the roofs off houses, burst the levadas, flooded the town and caused several shipwrecks in the harbour. Furniture and barrels of wine floated down the streets as people climbed on roofs to avoid drowning. The American brig *Creole*, the Jersey schooner *Wave*, the Portuguese schooner *Nova Beijenho* and the English trading vessel *Dart* were all dashed to pieces on the rocks.

Although this storm did immense damage to the crops it was nothing to the crisis ten years later which threatened complete extermination of the vines. This was due to a fungus blight which arrived in 1852. Hushed voices still speak of this and of the aphid attack in 1873 as though they were only a year or two ago.

Equally sudden and non-recurring was the cholera outbreak in 1856 which decimated the population. A festa in Funchal, the overflowing of drains and the arrival of the 1st Regiment of Infantry from Lisbon combined to start and spread the outbreak. Before the island was free of it on 30th October, Mr. Stoddart, the British Consul, reckoned that 8,000 people had died out of a total population of 105,856. The British themselves lost 15 out of 376. Mr. Stoddart had already collected generous contributions from the British residents before his wife became one of the victims herself. Another victim was the elderly Dr. Archibald Ross who heroically cut short his leave in England to open extra hospitals and treat the sick. Never did this ghastly killer return.

Madeira is the largest small island I know. Its size on a map, 35 by 13 miles, is most deceptive. There is just one stretch of road which is more or less straight for about three-quarters of a mile, but often one has to travel four miles to reach a spot visible one mile away on account of the precipitous gorges which have to be outflanked.

Strangely enough there are no hotels outside Funchal at present —only a few bare rest huts in which to camp—so the average visitor does no more than hire a car to drive to some beauty spot. Thanks to this the people outside Funchal live simply and retain their natural charm.

The country roads are fringed with hydrangeas and agapanthus lilies. Scarlet poinsettias, bright bougainvilleas, masses of begonias and geraniums, lovely yellow alamandas, pink or white oleanders, big white-trumpetted daturas, blue plumbagos and a variety of orange or pink bignonias greet one in every village. Madeirans cherish their gardens with loving care.

The cliff scenery of the sea roads is delightful but it is the wild and spectacular mountain scenery of the interior which leaves one spellbound. An early visit to see the Grand Curral from Pico do Sorrado left us panting for more. From this vantage point one looked almost vertically down 1,500 feet into a huge emerald valley dotted with tiny white houses on its floor. And looking upwards we had our first view of a range of high mountain peaks topped by Pico Ruivo (6,104 ft.).

Contrasts in this vast small island were provided by four miles of grassy savannah on the treeless plateau of the Paul da Serra at 4,690 feet, by the waterfalls and beautiful woods at Rabaçal where tree heather trunks sometimes reach a girth of six feet, and the virgin forests of Til (*Ocotea foetens*) and other Madeiran laurels in the gorges reached from St. Jorge.

The ascent of Pico Ruivo took place at the end of August 1964 when Jean and I combined with three naturalist friends, Eric Classey, Timothy Tams and Raymond Uffen. As Jean was neither a naturalist nor a mountaineering type, I felt proud of her determination to come, and although she did not enjoy the discomforts she emerged with honours.

Having booked the rest hut for two nights and arranged for a guide and bearer to meet us at Queimadas (2,952 ft.), we set out by car from Funchal.

The morning was unpromising with the hills covered in a

blanket of cloud which was entered at Monte and remained with us as we crossed the pass at Poiso (4,592 ft.) and descended into the lovely valley of Ribeiro Fria. Visibility improved near sea-level at Faial, but views were again blotted out as we mounted once more between Santana and Queimadas. At Queimadas we waited for the guide and bearer to arrive, outside a large house and garden.

It started to rain and all was silent except for the dismal sound of rain dripping from the trees. It was cold and we began to wonder if the guide would ever appear. 'How nice it would be,' Jean said, 'if somebody came out of that house and invited us inside for a cup of hot coffee.' The suggestion was greeted with mocking laughter by our companions, but Jean and I strolled up the drive and, hey presto, it was done. The other three could scarcely believe us when we walked back to tell them that a charming Portuguese lady had issued an immediate unsolicited invitation.

She turned out to be an army officer's wife staying there with her husband on leave and the gardens were a government horticultural establishment. For more than an hour she warmed us with coffee and conversation until the guide eventually arrived.

This guide was a small dark young man and he arrived alone except for his small barefooted eight-year-old son. When would the bearer arrive to help carry some of our loads, we asked? He pointed to his chubby little son!

The guide strung far more of our baggage on a pole than seemed possible, balanced it carefully on his shoulder, handed two baskets to the small boy and away we went with comparatively light loads ourselves.

The weather cleared as we walked up a dry rocky watercourse with high fern-covered banks. This continued almost interminably with the banks obscuring all view until it brought us to an open turf-covered shoulder of hill with a castle of rocks at its further end over which we clambered. Now that we had the chance of a view we found ourselves looking down on a white curtain of cloud pierced by jagged mountain peaks.

Presently the path led along the side of a bare ridge where huge boulders seemed to be poised waiting to roll down on us. At times this path had an annoying way of travelling downwards to lose some of the altitude we had gained, but eventually it came to the foot of an abrupt slope of loose rocks on whose summit our rest hut was perched a few hundred feet above us.

If Jean was tired at the end of our three-hour climb she concealed it manfully despite sorely bruised feet and was soon cooking a meal on an oven we stoked with tree-heather stumps. The hut comprised one large room with a fireplace, a table, and wooden chairs. Three doors on one side of it led into cells, the size of a railway sleeping compartment, each containing six bunks. Straw-filled mattresses and pillows with rough oily blankets completed the bedding. The primitive kitchen and, surprisingly enough, a water closet lay beyond the main room.

While Jean was busy cooking, the rest of us climbed the remaining 200 feet to the cairn at the top of Pico Ruivo (6,104 ft.) but the only view that evening was of mountain tips purpling in the evening light above the clouds fastened round their shoulders.

An early rise at dawn provided a gorgeous spectacle. The air was clear, except for thin clouds like smoke wisps rushing at speed between some of the lesser peaks. Brilliant orange and pink lights lit the mountains save those in shadow which were dark purple. Floating in the distant sea were both Porto Santo and the Desertas changing in colour from pink to gold as the sun rose.

After breakfast we all climbed a path fringed with tree heather to the little platform at the top of Pico Ruivo. Deep and mysterious valleys radiated outwards towards the sea. Most of the mountain summits had bare rocky crests rising above their green canopies of dense forest. Here and there thin waterfalls fell like white ribbons into the valleys beneath. To the west a bare razor-edged ridge curved downwards from Pico Ruivo and marked a sinuous route by which Bica da Canna (5,248 ft.) could be reached seven miles away along the island's backbone.

The rest of the day was spent making natural history collections. After dinner we stoked the fire and were playing a game with match boxes invented by Eric when peace was broken suddenly with the door being banged open to admit six noisy Portuguese who came bursting in with wine bottles in their hands. At first we thought they would leave again after a rest, but presently it became clear that they were intending to stay the night. This nearly created an international incident, but peace was restored when it was found that the hut had been hired wrongly to both parties for the same night! All five of us crammed into one of the cells and unfortunately it was the one in which fleas had been found during the previous night!

The naturalist who visits Madeira may at first be disappointed with what he finds. Whether he is interested in birds, wild-flowers or insects he may notice fewer species than he was expecting. He has to learn where to look and even then he must be content with quality rather than quantity. Anything up to a quarter of what he sees or collects live nowhere else in the world or else are shared only with the Canaries and Azores. Seemingly familiar birds, plants and insects turn out to be special sub-species. This can be illustrated by a moment of jubiliation in the hills close to St. Jorge when returning from a journey into dark mysterious valleys of virgin forest. A kind forester was reviving us with strawberries from his garden when Madeira's special brimstone butterfly was seen. Three swishes of a net resulted in the capture all at one time of this brimstone (*Gonepteryx cleopatra maderensis*), Madeira's special large white (*Pieris brassicae wollastoni*) and the island's own colour form of the red admiral.

Several of Madeira's endemic wild flowers are large and beautiful but their discovery in the hills was eased by our kind resident friends, Gerry and Irene Maul, and by Major Pickering. A mauve orchid, a thornless rose, a showy chrysanthemum, a blue echium, a sonchus thistle and many others cannot be seen anywhere else.

Madeira is an island to which most people want to return.

CHAPTER TEN

The Painted Desert

MADEIRA, Porto Santo and the three Desertas form a group of Atlantic islands separated by about twenty miles from each other. Considering their closeness they are surprisingly different. Madeira is lush and thickly forested; Porto Santo is open and edged with white sandy beaches; the Desertas are well-named being barren, waterless and uninhabited.

An intention to visit the Desertas someday dated from 1923 when I sketched their outline from a ship taking me to Brazil, but the opportunity never came until 1962. Two dreams made me tingle at the prospect.

One dream was of hidden treasure! This aroused romantic thoughts however remote the prospects of success might be. As a schoolboy I had read E. F. Knight's book, *The Cruise of the Alerte*, in which he recorded the story of the treasure from a fourth-hand account. A Spanish ship laden with two million dollars in gold and silver had reached Madeiran waters when news came of prowling British frigates. Spain was at war with Britain at this time, in 1804, so the crew decided to take desperate action. They killed their captain and buried him above the treasure in the sandy bay of an uninhabited island in the belief that anybody who found the body would dig no deeper. The island was the central one of a group. The mutineers then sailed westwards intending to purchase a fresh ship in which to return for the treasure but they were wrecked near Tobago with only one survivor who disclosed the secret before he also died.

If the story was true the unnamed island was in the centre of a group of uninhabited islands somewhere not many leagues from Madeira. Treasure hunters had scoured the Salvages without success so the Desertas obviously represented another possibility.

As for the other reason for wanting to stay on the Desertas, this also offered treasure of a different kind which would not appeal to most people. It was a giant spider—the largest Wolf Spider in the world—which was confined exclusively to the tiny island of Deserta Grande. When it was found there over a hundred years previously the discoverer had said that it fed on lizards. This was certainly possible for a spider with a body length of 4 centimetres but it would be a unique custom for a spider if it were true.

During the flight from England to Porto Santo a clear view showed the Desertas to be arranged in one thin sinuous line stretching across twelve miles of sea. Their appearance was suggestive of a sea serpent with a small flat head (Ilheu Chao), a long slim body crested with mountains (Deserta Grande) and a toothed tail whose tip was below the clear water (Bugio). All these islands were scorched and biscuit-coloured in the June sun.

It is nice to learn whatever is known about an island before reaching it, but the Desertas had very little recorded history. Fructuoso had written that Zargo introduced goats, sheep, cattle and domestic fowls in the 15th century. After his death the islands remained the property of his Goncalvez descendants until they were sold to Alexandre F. Camacho by the eleventh generation in 1870. All other property had had to be sold in the fourth generation to pay debts left by Simão Goncalvez da Camara, whose nickname Magnifico described his spendthrift way of life, so his son Luiz Goncalvez had done his best to make a profit from the islands' resources which were meagre. In his time there had been houses and a church. The goats were already described as being wild but besides having sheep, cows and rabbits he planted wheat and barley in the shallow soil. A track was made along Deserta Grande's crest and he constructed circular threshing floors and clay-lined ponds to trap rain and dew.

The most profitable ventures came from the sale of shearwaters and orchilla weed. Thousands of shearwaters were trapped for their flesh, oil and feathers. As for the orchilla weed (*Rocella tinctoria*), this was a lichen which was much in demand for the violet dye it produced and it was exported to Europe where the highest price of £90 per ton was reached early in the 19th century. This may sound a lot, but the task of tearing a ton of lichen from the rocks must have been a prodigious one.

A different strain of goats had been let loose on Bugio but only intrepid cragsmen could stalk them and no one could cultivate crops on so jagged an island. Cereals were harvested on the flat top of Ilheu Chao and ovens were installed for making soda from the barilha, or Ice Plant, (*Mesembryanthemum crystallinum* and *M. nodiflorum*).

Disaster came to the Desertas in the year of 'the Great Flood', 1802, when a terrible hurricane accompanied by water spouts damaged the houses and washed most of the top soil into the sea. This marked the end so far as resident population was concerned, and from that time onwards its solitude was disturbed only by sportsmen or those going to collect shearwaters and orchilla weed until the 1939–45 war. During this war some Germans were captured after being landed on Deserta Grande to signal news to their submarines and their places were taken by Observers who lived in two rough shelters and in caves fitted with bullet-proof doors.

Goats, sheep, rabbits and cats had survived in a wild state after the exodus of herdsmen but the last sheep was shot in 1852.

Camacho had coaxed a clump of pine trees to grow on the summit of Deserta Grande which throve long after he had sold the islands to two sportsmen, C. J. Cossart and H. C. Hinton in 1894, but these trees were used as firewood by the resident Observers during the war. To bring history up to date, the islands are again offered for sale at the time of writing (1968).

19th June 1962

The day started bleakly with bread, jam and coffee from thermos flasks at 2.45 a.m. An hour later we were assembled on the moonlit quay at Funchal waiting for Dr. Durão and Alec Zino who were going on a day's shooting expedition to Deserta Grande and had kindly agreed to give us a passage in their launch. My daughter Richenda and I were to camp there with a guide for four days while our friends Timothy Tams and Eric Classey were to be picked up on the following day.

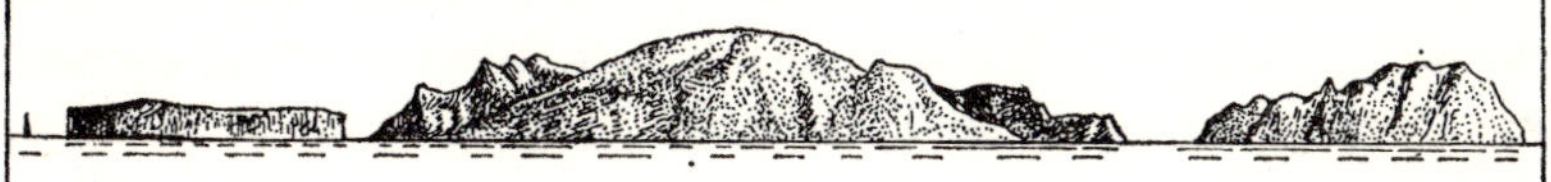

Desertas

Both our friends were naturalists and we could not have had nicer companions. Timothy at seventy had a benign appearance which amply expressed his kind and whimsical nature. His scalp, fringed with a snow-white fleece, had turned scarlet in the Madeiran sun to the delight of all the Portuguese children with whom he had made immediate friendships. Never had they seen such a flaming glory. He and Richenda had already established a close affinity of youthful outlook despite half a century's difference in their ages and neither concealed their glee when I warmly greeted as Dr. Durão one of his servants arriving on the quay with the doctor's gear.

Eric Classey was also a source of amusement to Richenda both on account of his quick schoolboyish wit and of his shrill excitement whenever he brandished his butterfly net in pursuit of some insect with a long latin name.

By 4 a.m. the *Espadarte* set sail. Sitting on the foredeck, we all watched the changing silhouette of Madeira and tried to identify its villages by their twinkling lights. Then, with the first rays of dawn, the fleeting forms of shearwaters skimming over the water could be seen and an occasional flying fish embarking on ghost-like flight over the rippling sea. In the morning light the three Desertas could be picked out. To the left lay Ilheu Chao with seemingly vertical sides and a flat top. In the middle Deserta Grande, lofty, precipitous and, from our angle of approach, deceptively shorter than six and a half miles in length. Beyond and to the right, Bugio with sharp toothlike peaks.

We were landed for two hours on Ilheu Chao while the shooting party were taken to the northern landing on Deserta Grande to start their trek to the southern end where we would meet them later. Time allowed us to scale the 350-foot cliff and traverse Ilheu Chao's rocky tableland. Faint indications of walls, threshing floors, dew ponds and ovens for burning barilha were the only signs of past habitation. The pretty flowers of the barilha, or ice plant, were now over, leaving a carpet of fleshy pink leaves. A large mauve thistle, the handsome blue *Echium plantagineum* and a white poppy (*Papaver somniferum nigrum*) were still lingering on though past their prime. An occasional painted lady or clouded yellow butterfly fluttered in the morning sun.

After rejoining the *Espadarte* she sailed along the western face of Deserta Grande. So steep and forbidding were the cliffs that boulders seemed likely to drop into the sea beside her and no safe ascent seemed possible even when we cast anchor in the shelter of a small islet. This islet, which conveniently provided shelter for the boat, had been formed by a gigantic landslide in 1891, the effect of which had been a tidal wave sweeping up the streets of villages in Madeira more than twenty miles away.

Without waiting for another breakfast a landing was made at 8.30 a.m. and our guide led us to some crumbly steps cut into the cliff. These led towards the main cliff but as yet no feasible route up its vertical face had revealed itself. Presently the steps led round a slight prominence to a steeply slanting path scratched out of the crumbling rock to a width of three feet. Hunger and thirst were forgotten as we toiled up and up in the hot sun with heavy loads on our backs. To practised climbers no feelings of fear would have been caused by a vertical wall on the left which crumbled at the touch of a finger and with one stumble to the right leading to eternity, but to me it was a terrifying experience as we mounted to 500 feet, to 1,000 feet and finally to 1,400 feet with a high dive straight into the sea still possible.

My gaze was concentrated on the path, while Richenda, seemingly untroubled, bounded on ahead with the guide. The irrepressible Eric remained as my bugbear with a flow of merry chatter which only a rare grunt in return did nothing to stem.

1. *Mt. Beerenberg from a distance of 25 miles*

2. *Hagbard Ekerold, the author and Tom Lethbridge on Jan Mayen Island*

3. *Eilean Tighe from Garbh Eilean, Shiar*

4. *A beehive cell and ancient cross on Skellig St. Michael*

5. *The* Viking *in full sai*

6. *On the Desertas*

7. *The Dalmatian island of Korcula*

8. *Ceylon's Sighiri fortress*

9. *Cave painting on Sighiri*

10. *Companions on Krakatoa*

Marie I, King of the Sedangs

12. *On a mushroom-like islet in the Philippines*

13. *Gannet colony on Grassholm*

Oblivious of my situation he exhorted me to admire the view or to notice past landslides as he kicked boulders over the edge to estimate the height from the time they took to reach the bottom. He even swished his butterfly net at a grasshopper, remarking that he had saved it from a record jump over the edge.

Meanwhile Timothy Tams was taking his time, so we waited at an inward curve in the path where it followed the course of a small gully. In one hand he carried a Gladstone bag and his butterfly net; in the other a precious lantern with which to entice moths at night; round his neck dangled a knapsack and camera. Frequently he stopped to take photographs or capture an insect, and his first action on joining us was to produce a spirit level, of all things, to see if at this point the path's gradient was slackening. This was typical of the mysterious contents of his knapsack.

'It's no use thinking of things you might want,' he explained, 'unless you always carry them with you.' True to this doctrine his knapsack contained a screwdriver, a composite knife, a ball of string, a leather-sewing needle, sticking plaster, spare shoe laces, some brandy and a piece of white chalk. When we finally reached our camping site it was noticeable that Richenda had to produce a tin opener and I a bottle opener!

Just below the topmost crags at 1,400 feet the path became level and widened into a 7-foot platform in front of the two small caves fitted with bullet proof doors by the wartime Observers. This was to be our camping site though we made little use of the cave except as a storeroom where the doors could be used to exclude a multitude of lizards from devouring our provisions. Both here and in Madeira these lizards were omnivorous and would wrangle fiercely over tomatoes or cake just as keenly as if it had been meat or an insect.

Books say that Deserta Grande has no water but Juan de Sã, our guide, had found a tiny trickle in a cliff face, across the island, and had cunningly led it by a runnel into a storage tank the size of a suitcase. While he fetched water, we changed our clothes and then Richenda cooked us a meal.

The path beyond the camp led into a small valley of red earth and lichen-covered rocks with patches of close-grazed dry grass.

A pair of fawn-coloured goats with black markings along the crest of their bodies were visible 200 yards away and they saw to it that the gap was never narrowed. An occasional Berthelot's Pipit (*Anthus bertheloti maderensis*) flew from rock to rock, and once or twice greenish canaries were seen near the cliff edge, but otherwise all seemed still and lifeless. The turning of rocks produced nothing besides millipedes, cockroaches, beetles and small spiders.

The next valley had no vegetation at all so we circled back and started in a northerly direction. After climbing steeply from the camp we came to a valley of brown earth baked into wide octagonal cracks, and from a high point on its further side a good impression of the island's narrowness was provided. From here Deserta Grande looked like part of the rim of some gigantic extinct volcano.

Beyond this valley lay another with flame-coloured pinnacles of rock and a soil streaked with vivid ochre. And then in succession came others where the soils or rocks were slate blue, grey, black, white, yellow or dull red, until nearly every colour in the spectrum had been exhausted. In some valleys there were weird toothlike pinnacles and in others large rounded nodules peeling in layers from a central core which were reminiscent of giant onions (Pl. 6).

This strange lunar landscape, this painted desert, had a weird fascination of its own and, coming straight from nearby Madeira with its flowers and forests, the contrast was extraordinary.

Walking along the island's crest we came presently to a loose precipice and a continuing knife edge with sharply sloping sides which completely blocked further progress. This spot had obviously been the site of an extensive landslide on the other side of the island. Up till now we had expected to meet the shooting party but they had used a route, which Richenda and I were to follow later, that dipped down nearer to sea level on the other side of the island and up again to the summit not very far from our camp. On our return from this reconnaissance we were in time to bid the shooting party farewell with their bag of three goats before preparing our evening meal.

During the meal somebody began a discussion on what we should do if the treasure were found. Could we all be trusted not to cut each others' throats, for instance, and what about the boatman who came to fetch us? Was there a chance it would be confiscated in Madeira and what would happen if we got it as far as the Customs in England? Could we possibly keep its discovery a secret until we could return in a chartered boat manned with trusted friends? Could anybody be trusted once they had set eyes on millions of pounds worth of doubloons? Was there any country to which we could convey it and feel sure of retaining the lot? Simulated feelings of greed, suspicion and roguery made the discussion an amusing one though we all knew by this time that there were no sandy bays on Deserta Grande where the treasure could have been buried. Cliffs diving steeply into the sea did not answer the description at all.

Our guide, Juan de Sã, was a peasant of indeterminate age with a gaunt lined face that lit into a delightful smile at the slightest suspicion of a joke. He came from a long line of herdsmen from the village of Canico who had been connected with the Desertas for centuries. We never saw him, night or day, without his knitted brimless wool cap with its side flaps and thick flared top knot. He liked to brood in solitude sitting on a rock at the very edge of the precipice like some huge bird of prey. Whenever we joined him to ask a question he repeated his answers over and over again in lilting Portuguese until we had decided what we thought he meant.

This part of the island was obviously too bare and arid for the giant spider so Richenda accepted the task of questioning Juan de Sã with the help of a Portuguese dictionary. From this discussion she learned that a valley of long grass lay at the extreme northern end some six miles distant from the camp, that the way was difficult and that it would take two and a half hours to reach. Making allowances for collecting on arrival and the return journey down to the harbour, it was clearly impossible for Timothy and Eric to avoid the risk of missing their launch which was booked to collect them in the mid-afternoon of the next day. This was a bitter blow because both of them had become keen on the spider project but it was slightly softened by the promise of live specimens and champagne if success was achieved.

The island was very quiet after they had gone. Juan de Sã sat brooding on a rock where we joined him to watch the faint cone of disturbance behind the launch as it disappeared into the distance. Presently our attention was distracted by a school of whales disporting themselves and later the evening glow converted the near-by cliff face into a sheet of dazzling gold. A cloud began to descend on the distant mountain tops of Madeira as we spread our bedding on the ledge in front of the caves and prepared for an early night's sleep. Here we lay looking up at the stars with our toes close to the cliff edge and nothing between them and America except thousands of miles of sea and sky.

Hopes of the day ahead were just blending into a dream when a sudden swish close to our heads startled us into wakefulness. The black forms of cagarras or Atlantic Shearwaters (*Puffinus diomedia borealis*) could just be seen as they skimmed within feet of our heads like giant swallows. Presently some gurgling noises below us developed into sounds like the barking of dogs and the moaning of children. This brought eerie responses from the birds in flight, screeches which gathered or diminished in volume according to the direction of their flight. Ung, ung, ung, awww, they wailed which seemed hardly a reassuring message for their families at home. Perhaps it was the rising moon which silenced them at last or else we fell asleep with our heads under the blankets to lessen the din.

Waking before sunrise with the intention of an early start, we were dismayed to find ourselves enveloped in thick wet cloud. Breakfast was prepared but even in daylight there was no view beyond the ledge on which we were perched. Juan de Sã took a gloomy view and said it would be impossible in this weather to lead us to what we already called Spider Valley, but eventually a slight lifting of the cloud, combined with Richenda's coaxing and my persistence, enabled a start to be made at 7 a.m.

A steep ascent to the central saddle was soon followed by an equally sharp descent on the other side to a horizontal path cut out of an almost vertical cliff. In places I had to stoop to avoid hitting my head on the overhang. For nearly a mile we walked along this narrow ledge with a white cauldron of mist shielding us from a view of the drop below us. Twice we came to places where the three-foot path had been swept aside by landslides and even Juan de Sã walked as delicately as Agag using his iron-tipped fifteen-foot pole to test the ground ahead of him at every step.

At the end of the mile this cliff path curved inwards and over the island's saddle to a relatively wide firm path at the cliff edge on the other side. Presently this led back on to the central core where it vanished, leaving Juan de Sã to conduct us through some amazing lunar landscapes. Bare valleys with high pinnacles of rock, bright ochre in one and flame-coloured in the next, looking all the more weird through the mist. A valley of large white boulders looking like tombstones on a dark grey floor suddenly gave way to a patch of undulating moorland where grass and thistles glistened with drops of moisture, or occasional clumps of the blue *Echium* alongside the unique Deserta sowthistle (*Sonchus oleraceus laciniatus*) livened the scene with colour.

Beyond the moorland we emerged once more into desert and presently we found ourselves slithering down loose red soil towards an obscured cliff edge. A retreat to higher ground then brought the party safely to a different slope which led down into a smooth-sided valley of high grass and herbs with a small ruined Observer's house in its centre.

We had arrived! At last the terrain looked favourable and as this was the one and only corner where the giant spider could survive on the island where it had been created, it must surely be here in abundance. In great excitement we gently tilted a large boulder and stood spellbound with excitement. There were two of them, a male and a female!

To me, of course, this was a memorable moment as I had longed to rediscover this spider for over forty-five years. In admiration I gazed at it in silence, taking in its appearance which had never been illustrated or adequately described. Part of my mind was also picturing how differently Timothy and Eric would have reacted and envying them for their greater ability to show their feelings. I felt sure they would have shouted and skipped with delight yet here was I merely staring admiringly at the huge spider with its startling white rings and spots on black legs raised in an attitude of menace. Large shining black jaws were bared. The body itself was grey. It was indeed a superb and defiant prize.

The trance was broken by the male darting for cover. Out came a net and the female was forced, fighting every inch, to run into it for transfer later to a polythene bag.

Every third or fourth boulder had this Wolf Spider, *Lycosa ingens*, underneath it. Each female lived in a roomy silk cell into the walls of which the remnants of its food were woven. Although lizards were plentiful the spiders did not prey on them, but the original discoverer had probably reached his conclusion from the many bleached rings of millipede chitin which he mistook for lizard vertebrae. Other remnants proved that large beetles and cockroaches formed an important part of their diet.

By the time five males and fifteen females had been ushered into polythene bags, Juan de Sã was getting restive so we made our way home in swirling mist and rain showers. It was not until camp was reached that I realised the dreaded path had made no impact during the return journey and that I had even stopped to take photographs. Subsequent travels have proved that these days on the Desertas have evaporated my lifelong fear of heights.

Postscript. Madeira, Porto Santo and Deserta Grande in their different climates have each evolved a giant species of Wolf Spider from some common ancestor but, strangely enough, the size is in inverse order to that of the islands themselves. Though closely related they are easily distinguished in the living state. *Lycosa ingens*, from Deserta Grande, is black, white and grey. *Lycosa maderiana*, from Porto Santo, is paler in colour and has striking orange legs. *Lycosa blackwalli*, from Madeira, is more like *L. ingens* except for some scarlet markings on the legs and black chevrons on the abdomen. The habits have slightly diverged also. *L. ingens* lives in silk cells, *L. maderiana* lives free, and *L. blackwalli* lives in short burrows under stones.

When preserved in spirit these closely related species lose their bright colours and are not so readily distinguishable at a glance. This has led, I believe, to *L. ingens* and *L. maderiana* being incorrectly recorded from Madeira.

The Giant Wolf Spider

CHAPTER ELEVEN

Commentary on the Mediterranean

THE Mediterranean is blessed with lovely islands of which fragrant memories in the 1930s are sweeter by far than any return visit could provide. Climate, historic interest and much of the beauty remain, but to an island-lover seeking peace and romance or adventure and the unusual, the atmosphere has undergone the inevitable change which popularity brings to man and island alike.

Concrete hotels are springing up like mushrooms on Malta and even on its little satellite of Gozo. Sicily has lost its bandits and there are dense pockets of sophistication with package tours, cable cars up Etna and a welter of loud visitors. Thoughtful hotel managers even provide free turpentine for guests using their beaches in order to remove the slicks of tarry oil inadvertently collected by swimmers and sunbathers alike.

The Venetian islands had, of course, always been a beehive of activity but it used to be quiet on the canals except for the leisurely gondoliers who enhanced the romantic settings by singing as spontaneously as larks when their poles drifted one smoothly through deep shadows into the bright sunshine beyond. Those who remain give little and extort much as they steer their way through the fumes, noise and wash of countless motorboats hastening to some unknown destination. History and delightful architecture remain, but so do nostalgic memories for those who knew these islands before the crowds of holiday makers and motor boats invaded the tranquillity of the waterways.

Millions of visitors now throng annually to Majorca where in 1930 the few were welcomed both by friendly islanders and the small resident community of English people who had retired there to enjoy its beauty and cheapness. Where are the innumerable groves of incredibly ancient olive trees which demanded a pencil to sketch their trunks contorted into weird shapes like wrestlers sparring with their neighbours? No longer can the hospitable monks of Lluch monastery offer to pilgrims the free accommodation I once enjoyed in their mountain fastness for a period of peaceful contemplation. Luxury hotels have sprung up which charge daily what their adequate predecessors expected in a week.

No one should complain that crowds of people can share the pleasures of an island holiday but it results in a person with different tastes having to search for what he wants elsewhere. It means also that experiences of the 1930s are mostly too outdated to be written about.

The Greek islands still have to be revisited though nothing could reproduce the magic of a cruise with Stephen and Ginie Courtauld in 1933. Part of the wonder of that cruise was to see how the atmosphere of the islands combined with our hosts' charm to provide complete harmony between their guests who might have been selected by a writer of thrillers as a prelude to dark tragedy. The Courtauld genius had restricted their guests in the *Virginia* to people greatly in need of just such a holiday. They comprised a temperamental impresario, a famous opera singer who had suddenly lost her voice for ever, a husband and wife experiencing the difficulty of one becoming an avid convert to the Oxford Group during its most embarrassing stage of open confessions in church, a girl whose fiancé had just been drowned during an Arctic expedition, two ordinary young men who could never have afforded such a holiday on their own, and the charming girl friend of one of them.

Difficulties melted and cares faded into the background, or almost. At one Italian port of call I had remained on board and could hear the opera singer, Margaret Sheridan, playing some gramophone records of her own singing in the seclusion of her cabin and trying without any success to imitate them. A crowd of people collected on the quay to listen. Presently she came on deck. Noticing the crowd, and realizing they had been listening, she walked to the ship's rail and asked one young man if he knew whose voice he had been hearing. 'The Sheridan' he said in a voice of reverence. Poor Margaret was so overcome that she turned away speechless, and retired to her cabin once more with tears in her eyes.

From Brindisi we sailed for the islands which mirror history since the dawn of civilisation, the homes of gods, giants and heroes, the resort of biblical saints and crusaders, the site of wars, peace and beauty. Unfortunately, they deserve so much more than one superficial gleaning that I am forced by true humility to pass them by save for a few fragmentary impressions.

Multitudes of spired cypresses gave greenery to the slopes of Corfu as we sailed in view of its citadel and ancient Venetian fortress. Stephen's erudition brought home to me the wasted years at school when boys were forced to translate or learn by heart small fragments of Homer and Virgil without ever getting the sense of the complete stories. However, it was amusing to learn that we were bathing in the very spot where Ulysses had played ball with Nausicaa and her lovely maidens!

The smell of thyme and sage reached across the sea as *Virginia* drew near to Ulysses's island of Ithaca whence he had been summoned to the Trojan war despite his attempt to feign madness by ploughing with a team of strange animals and sowing salt instead of barley. It was he, according to Homer, who contributed to success by inventing the famous wooden horse, but he was away for so long that his wife, Penelope, did not recognise him on his return until he had enacted his special parlour trick of drawing his bow and shooting an arrow through twelve rings.

By comparison the drier Cyclades group of islands seemed to glow with a golden light reflected from their sered grass and to float lightly on the deep blue water of the Aegean sea. Each in turn seemed more interesting and beautiful than the last. Double peaked Melos had yielded the head of Asclepius to the British Museum and the Venus of Melos to the Louvre. The perpendicular crater walls of volcanic Santorin and the ancient capital of Oea where the earliest Greek alphabet was punched into the bare rock both attracted us to its shores. Mykonos, with hill ridges thickly crested with windmills, was richly endowed with numerous Byzantine chapels. And finally Delos, the sacred isle, whose wealth of history and ancient sites deserved intimate study and infinite imagination.

Close to the landing stage was the columned portico built by Philip V of Macedon over 2,000 years ago which was modern compared with some of the other relics. A terrace some way beyond it held the foreign gods of Egypt and the sacred precinct itself, the remains of ancient temples supposedly built for Apollo, Leto and Artemis.

A rocky path led up Mount Cynthos from whose summit small golden islands of differing shapes and sizes could be seen studding the sea in every direction. Below the summit is the earliest shrine of all. This goes back to prehistoric times and is in the form of a simple cave artificially arched by two huge stones leaning inwards against one another.

The sanctity of Delos is attributed by Homer to Leto who persuaded her son Apollo to erect a temple after she had found refuge there. For long ages it remained as the religious centre of the Ionian states with new temples and fine buildings being erected throughout the Hellenistic period until the coming of the Romans who dulled the island's supreme sanctity.

Eastward to the Sporades *Virginia* came to rest close to the inlet which cuts deep into Patmos. Smiling faces and flowers from the children left happy memories of this little island to which the apostle John had been exiled by the Romans. Here in a simple cave fringed with moss and ferns he is said to have written his Book of *Revelation*. I know that I gazed in amazement trying to picture him, or anybody for that matter, writing a major work in such a place of discomfort.

Lastly we came to the Dodecanese and in some respects the most striking of the islands. Spurning the hire of donkeys for the ascent of Rhodes' grand hill slopes we toiled upwards to the famous rock sculpture of a galley at Lindos and also climbed the ruined stone steps up which Alexander the Great is said to have mounted. At a much later date they were used by the Knights of St. John who built a gate house at their head leading to their fortress. During their period of occupation in the 14th or the

15th century the crusaders built massive battlements, a hospital and many houses which survive in good condition, but one can go back much earlier to the ancient inscriptions and ruined temples where, as usual, keen imagination is needed to picture them in their ancient glory.

Rhodes is the island of the sun god, created specially for him, it is said, by Zeus. The Colossus of Rhodes, measuring 105 feet, represented him near the harbour and for a short period, it ranked as one of the seven wonders of the ancient world until it was shaken to pieces by an earthquake in the 4th century B.C.

Ruins and history can be encountered in plenty but, quite apart from these, Rhodes stands high amongst all the Mediterranean islands for its views and its beauty.

No detrimental comparisons with past visits could be made by Jean or myself when we visited Yugoslavia in 1967. Perhaps it was the stifling heat of July which at first failed to inspire us with sights of the country's thousand islands as we traversed the coast road between Dubrovnik and Split. The larger islands seemed to flow continuously down the coastal waters like shoals of long dark green fish with little sign of buildings or human habitation. Some of the lesser isles were bare and wrinkled like moonscapes gleaming white in the sun, their furrows buried in dark shadows.

Occasionally I sat up to take notice of the guide's intermittent remarks. 'That,' he said, on one occasion, pointing to a minute islet near Split bearing a solitary house, 'is called the Island of Love in contrast to that other one,' pointing to another with half its area occupied by six houses in a cluster, 'which is called the Island of Life.'

Both were attractive and in my imagination I began choosing my companions for a life on each!

A few days later the breathless heat of Dubrovnik was briefly relieved by a breeze of our own making as we crossed the narrow channel to the small island of Lokrum to which we had watched others crossing during the daytime, and also at night silently in unlit small boats. Like many other of the islands a narrow line of white rocks just above sea-level gave way to a dense forest of evergreen trees whose canopies were pierced by the dark pointed spires of cypresses. On the island's summit a pole marked the remains of a small chapel built by Richard Cœur de Lion in thanks for the sanctuary from storm provided by Lokrum during his homeward passage from the Crusades in the 12th century. Here this knight errant's name survived in a peaceful setting though in Syria it had become a bogy with which to frighten recalcitrant children.

The Yugoslavs are intensely proud of their country and loyal to their own particular districts so perhaps our blue-eyed young boatman's response to our description of the Island of Love near Split was understandable. 'Lokrum,' he said, 'is the Island of Love. There is no other Island of Love in Yugoslavia. Anybody who spoke of another was telling you lies.' And then, in order to emphasise Lokrum's unique status, he repeated his vehement denial of what our guide had told us. With a mixture of scorn and ridicule, he went on to say that some islands further north were put aside for nudist visitors, mostly German and French, but Lokrum's popular name came from its beauty and shaded groves where wooing couples sought solitude and bliss.

From a small landing stage we walked in the aroma of hot resinous pine trees to an assemblage of stone buildings some of which had been a 15th-century monastery and another the country house of Prince Maximilian of Austria in the 19th century before he strangely became Emperor of Mexico. Here, in a garden scented with rosemary and oleanders, we sought shade from an acacia tree at a small table and watched local fish being cooked for us on a charcoal grill while crickets chirped merrily all round.

The larger islands were saturated with fragrant myths, blood-stained history and exquisite relics from the past which echoed the history of the mainland to a remarkable extent.

Neolithic man left flint knives on Korcula and legend was later to link the names of Antenor and Aeneas with the foundation of its principal town. The exciting discovery of an inscribed stone from the 4th century B.C. actually gives the names of Greek settlers with the areas of land allotted to them. How surprised these humble peasants would have been to know that they would be remembered more than 2,000 years later.

New masters over the Illyrian people were to arrive in Korcula during the Roman period and, after long years of stout resistance to their will, Emperor Augustus is said to have massacred or deported most of the inhabitants in 33 B.C. Roman rule over Korcula and the mainland was to continue for a long time and the Emperor Diocletian's interest in the district is still witnessed by the magnificent palace-fortress he built in nearby Split some 300 years later.

After the Roman Empire's decline Korcula came under the influence of Byzantium until Slav (Croatian) invaders moved in who have not since been dislodged despite many further conquerors. The Venetians flowed in and out on four occasions and so did the Austrians. The Turks did battle with them and Korcula was to be controlled in turn by Napoleonic France, by Russia, and even by Great Britain for a matter of months in 1815 when their commander, Peter Lowen, left as his memorial the attractive semi-circular terrace above the beach.

Korcula passed back to Austria under the terms of the Vienna Congress in 1815 and remained in their hands until the Italians occupied it between 1918 and 1921 after which the new country of Yugoslavia was welded together.

The island was still to suffer occupations and bombardments during the 1939–45 war, first by the Italians and then by the Germans, but the Croats had succeeded in establishing such close knit municipalities and trade brotherhoods that they never seem to have been fully subjugated since their arrival in the 9th century. From the earliest time they had exploited their excellent stone quarries, developed their artistic skills in masonry, cut their timber and built ships uninterrupted by their nominal governors.

The approach to the old town of Korcula is magnificent through seas the colour of lapis lazuli as the ship turns in a half circle round the low perpendicular cliffs of its jutting peninsula. Old stone houses roofed with light pink tiles pile upwards above the cliff top where a massive wall and rounded forts protect them from attack (Pl. 7).

The layout of the town is somewhat like the backbone of a fish with one main way travelling north and south joined by about ten narrow parallel lanes slanting to either side. None is wide enough for a motorcar, but all are convenient for overhead washing lines. Even on the hottest day a blessed cool air streams up these narrow lanes as one climbs to the small central square. Bearing in mind that Korcula is only one of dozens of islands deserving a visit along this coastline the feast awaiting any visitor interested in architecture or the arts is amazing, a high proportion of the best dating from the 14th to the 16th centuries.

Here for the first time are heard the names of local masons, sculptors and painters whose handiwork vies with that of imported masters from Italy.

The Cathedral provides an excellent example of this. The 15th-century portal by Bonina Da Milano is striking, but so too are the masterpieces of the local mason Marko Andrijic. The Annunciation by Tintoretto must be hailed as a great work, but so must be admired the ceiling panel painted by an unknown local artist in 1420. It also contains a fine icon of the 13th century from Crete, the Badijska Madonna, but the marble altar, the wooden statues of St. Roccas, the Virgin and Child and others by islanders such as Marko Milic Pavlocic and Frano Cucic are not outclassed in such company.

The small square dominated by the Cathedral is full of 15th- and 16th-century stone buildings of merit and several of them are storehouses or museums of local or imported treasures from past centuries. Beside the Cathedral stands the 14th-century Bishop's Palace to which a beautiful 17th-century façade has been added. This houses a rich collection of Italian renaissance art in which Bellini, Porderone and many other artists are represented. Beside them are hung early Dalmatian paintings dating back to 1420.

Close to the other side of the Cathedral lies the more ancient church of Saint Peter beside the ruins of Gothic palaces and the claimed birthplace of Marco Polo, the famous navigator, who moved to Venice when still a child.

On the opposite side of the square are the Palaces of Gabrielli,

Ismaeli and Arneri, each containing its special treasures and providing impressive doorways, courtyards, balconies and staircases.

This and much more rewards the visitor while the countryside with avenues of pointed cypresses, the beaches and the friendly people all combine to tempt me back. To end on a strange note: Korcula is one of the few remaining places in Europe where jackals still survive. Interesting but scarcely an additional attraction!

My dominant impression of Yugoslavia was that this is a country whose works of art, not yet fully discovered, could inspire artists or antiquarians and after my return to England I was to learn that this is precisely what it had done to the Adam brothers in the 18th century.

The influence of Robert Adam's study of Diocletian's Palace in Split can be found, it is said, in the construction and interior creation of such buildings as the Adelphi (demolished in 1937), Luton Hoo, Syon House, Bowood, Kedleston Hall and Osterley Park. His illustrations of the Palace and his collaboration with Clerisseau and others also helped to spread this inspiration in France, Germany, Russia, Sweden and the United States. This influence was not confined to architecture. It extended to painters like Laurent Benoit, Dewez, and Agostino Brunias and to Italian engravers like Francisco Bartolozzi and Antonio Zucchi.

CHAPTER TWELVE

Of Monasteries and Prisons

THE isolation of small islands has been sought by holy men and thrust on the unholy. In both cases this isolation is related to the world's temptations. Monks remove themselves from the outside world lest they fall to its temptations while criminals are removed after falling to them. Both live in cells under strict discipline, but beyond this the similarity collapses because the one establishment is nurturing the soul whilst the other is destroying it for prisoners and guards alike.

The monastery breathes an atmosphere of peace and beneficence while despair or brutish cruelty is too often to be found in an island prison. The first would not suit me for long and the second would be a nightmare in the midst of beauty.

A few days at Lluch monastery in the mountains of Majorca acted as a wonderful tonic, but the life would soon have palled in the absence of the religious convictions held by the monks.

The homemade wine and splendid library in a monastery on Mount Parnassus left grateful memories of a Greek abbot, but no desire for the fate he had chosen.

Daily visits to the Lerins islands from Antibes provided some insight into both monastery and prison life. The island of St. Honorat was known as Lerina in distant times when the Romans anchored their ships in its lee, but it now bears the name of the holy man who settled there in the 4th century. The fame of St. Honorat, and one of his earliest companions, called Caprais, soon attracted numerous followers with a thirst for perfection. A sacred spring with healing properties was opened and the poisonous snakes for which the island had been renowned were all exterminated.

The claim is made that St. Patrick spent time there before being sent to Ireland at the bidding of Pope Celestine, so the special mention of snakes may be significant. Did he, perhaps, take part in the extermination of snakes on St. Honorat and did this give rise to his reputation as a snake-exterminator in Ireland where snakes never existed? It certainly seems possible.

A monastery and seven chapels were built among the pine groves of St. Honorat and the island was to gain fame as a seat of religious learning from which bishops, saints and martyrs appeared in profusion. It is strange to find how many links are claimed between this tiny island and early evangelism in Britain. Besides St. Patrick, there were Saint Loup of Troyes, who was sent to quell the shocking heresy of Pelagius that we were not all tainted with Adam's sin, and Saint Eucherius, Saint Hilary and Saint Cesarius all of whom came to England after a spell in St. Honorat. So, too, it is claimed, did Benedict Bishop of Northumbria and St. Augustine of Canterbury.

Much that is ancient can still be seen, including a 5th-century altar and the conspicuous 11th-century monastic fortress washed by the waves on three sides. An 8th-century attack by Saracens led to the martyrdom of no less than 500 monks, so a need for a fortified retreat had become painfully obvious, but its completion seems to have affected the monks' spiritual outlook because from that time onwards the lustre of St. Honorat departed.

Cistercians, who arrived eight centuries later in their brown or white cowls, are to be seen tending their crops and livestock, building, renovating and making green chartreuse, but a first encounter with them does not reveal their spartan way of life.

They talk for three-quarters of an hour on Sundays but otherwise obey the rule of silence. They rise every morning for matins and lauds at 3.10 a.m. and return to their cells at 8 p.m. after mornings and evenings of almost continuous religious services or prayers with short breaks for food. Masses, silent prayer, spiritual reading, matins, lauds, nones, vespers and nocturns follow one another in succession. Pious-minded people are not invariably good, but the long routine must help to keep them out of mischief.

Lerins: fortified monastery

Fereal, a tiny low-lying islet to the east of St. Honorat, was the site selected by Paganini for his own grave. After a turbulent life he lies here in solitude with the blue mountains of Estoril in the distance. His choice must have been made on a fine day, when the dazzling white convolutions of the sea lily (*Pancratium maritimum*) were in bloom because it is nothing but a dreary reef exposed to every wind and swarming with rats which emerge from its prickly scrub at night.

St. Marguerite lies to the west of St. Honorat, an island which once gained notoriety for its prison. For just under a quarter of a century it held the Man in the Iron Mask. Not even his warders knew his identity or crime and the cruel mask concealed his appearance. Louis XIV, who combined ruthless cruelty with an astounding ability to create beauty, alone knew the identity of his victim and this he never revealed except on his death to the Duc d'Orleans. The secret was passed on to Louis XV and Louis XVI, but it died with the latter on the scaffold.

The identity of this man of obvious importance to Louis XIV has provoked wild speculation. Was he the Duke of Monmouth, the Duc de Beaufort, Avedik the Armenian patriarch, or Louis's disgraced Minister of Finance, Nicolas Fouquet? Another school of thought believed him to be a relation of Louis XIV—his illegitimate son, Louis de Vermandois, who was certainly imprisoned for boxing the Dauphin's ears, or a twin brother of Louis XIV, or an illegitimate brother, or even, according to Dutch sources, a foreign chamberlain of Queen Anne, thought to be Louis XIV's real father. This hunt for candidates produced a fantastic list, many of whom could be certified to have died in other circumstances, but one candidate favoured by many was Count Ercolo Antonio Matthioli, a senator from Mantua, who had accepted bribes from others after negotiating a secret treaty under the terms of which Louis was to gain possession of the strategically important Italian fortress at Casale. It was thought that the infamy of the transaction was enough reason to hush it up and the fraudulent deception more than enough to cause Louis to be merciless.

The horror of this victim's existence for twenty-four years in a rusty mask is magnified by the sight of the dark dank dungeon in which he was incarcerated on St. Marguerite.

In Greek waters Nauplia is a grand sight from the sea with its ancient turretted fortress perched high and reached by dragging one's feet up nearly 900 steps. Roaming amongst its dungeons one notices massive bolted doors with the woodwork near ground level partly worn away by the kicks of desperate men. Heavy iron balls, which used to be attached to their ankles by chains, lie rusting in the corners and lizards run on the dim-lit walls.

Near at hand is a small islet completely enclosed by the walls of a Venetian castle which was used as a prison for the most dangerous criminals at the time of my visit in 1933. A provincial executioner, himself a convicted murderer, had been placed in charge some years earlier in the belief, presumably, that his homicidal instincts would be of definite advantage. He had complained of loneliness during a slack period so a murderer was sent to keep him company with the promise of more to follow. The promise was kept, but the island seemed deserted so a search party was sent ashore. It did not take long to find the explanation. Two terribly mutilated dead bodies pointed to a fight which had been fatal for both of them.

The idea of two murderers, alone on a tiny island, stalking one another night and day, is a ghastly thought. One of them had to die, but, in the event, both did.

The history of Mogador, now called Essaouira, off the Atlantic coast of Morocco is shared with that of the small island of Phraoum which acts as a shield to its harbour. Essaouira's attractive battlements with overhanging turrets were designed by a captive French engineer called Cornut for Sidi Mohammed ben Abdullah in 1769. The port was bombarded in 1844 by the Prince of Joinville, King Louis Philippe's brother, after Morocco had sided with Algeria against the French and given sanctuary to Abd el Kader. Finding the port too formidable to capture, the Prince put a garrison of 500 men ashore on Phraoum to intimidate the Sultan and a French victory in Algeria led to the siege of Essaouira being successful seven months later.

Nagel's *Travel Guide* had told us that the harbour was 'always calm' owing to the shelter provided by Phraoum, but I was to watch curiously red-stained waves venting their fury on the sea wall for four consecutive days in February before a boat could make a crossing to the island.

Seven old forts had been posted to protect every approach to Essaouira, each of semi-circular design comprising a thick wall, a terrace for five cannon, a well and two living quarters.

At the southern end was the lofty tower of a mosque which served as a look-out for approaching vessels.

Elsewhere along the cliffs the much earlier walls and relics of Roman and Etruscan occupation invited excavation.

Attempts to reconstruct the history of this not unattractive island with its views of Essaouira across the harbour had filled my thoughts until the remains of an old Moorish prison were reached. A massive twenty-foot wall surrounded an acre of open space with a watch tower from which armed guards could have seen every inch of the prison compound. Rusty metal bars built into the walls of darkened cells and the surrounding wall showed only too vividly where prisoners were chained by wrists extended above their heads in the cells or in blazing sun according to their guards' pleasure. Floggings and death from malnutrition or disease in the secrecy of these walls were probably part of the daily experience for the guards, grown callous with repetition.

Thoughts of man's cruelty oppressed me and even the cries of seabirds took on a note of melancholy as I walked back to the boat.

CHAPTER THIRTEEN

Ceylon's Sacred Tooth

CEYLON is too large and well-chronicled an island to claim much space in this book, but its natural beauties and magnificent lost cities recommend it as a first introduction to the splendours of the tropical East.

Leaving Colombo and the sandy beaches of Mt. Lavinia, the most popular route leads up into the hills past the wonderful gardens at Peradeniya to Kandy and the hill station at Nuwera Eliya where I have seen the eyes of a Scots companion light up with memories of his homeland as he roamed the grassy golf course in chilled mist. The heat and ample water supply of the plains near Colombo provides a picture of luxuriant vegetation and of the sleek brown skins of the villagers, so different from the parched soils and emaciated inhabitants too frequently seen in India. Onwards in an upward climb the road passes through serried ranks of rubber trees and impenetrable forests with gay flowers and butterflies on their fringes, to the neat tea gardens between Kandy and Nuwera Eliya where girls sing as they pluck the precious young leaves with trained dexterity.

Other roads lead to magnificent cities of the past choked by the forests which had besieged and concealed them for many centuries. Huge domed temples (dagobas) house statues of Buddha and provide fine sculptured elephants or scenes from the lives and hopes of past civilizations. There is much to admire but it is Buddha's tooth in Kandy which is the island's most sacred possession.

Once a year pilgrims in their thousands spend upwards of a week paying homage to a tooth they never see. At this time there are purification ceremonies in the lake, processions of richly decorated elephants, dancing to the point of ecstasy, stretches of burning charcoal for barefoot believers to walk across, and numerous entranced devotees with knives or spears thrust through their tongues or deep into their flesh. With sins purged and faith revived according to their individual needs, they can return home to a new year.

Faith is necessary in all religions and the early Christians were not different in their reverence of saints' fragments gathered by human vultures after their death. Nor do I believe that we are so different today when one thinks of the probable scramble which would take place amongst fans for relics of their favourite film or pop stars. A hair from the left eyebrow of such a star would be cherished far more highly than a locket enclosing grandpa's first curls. It is a hair from Buddha's left eyebrow that brings reverence to a domed temple at El Vehera and his collarbone to another at Thupatama. Likewise a picturesque Bo-tree at Anuradhapura is worshipped in the belief that it has grown from a seed planted in 288 B.C. from the tree in India beneath which Buddha used to sit in meditation. The thin canopy which surmounts a gnarled trunk is kept green in dry spells by moistening the roots with milk.

Faith has to be strained in order to accept that his much-travelled tooth is genuine. It was brought from India hundreds of years after Buddha's death two and a half millennia ago. It became a talisman, a symbol of kingship, so the ruling king had to carry it from Anuradhapura to Polonnaruwa in the 8th century and from Polonnaruwa to Kandy in the 13th century when the Tamil invasions made a change of capital necessary.

After arrival in Kandy it was looted by the Portuguese on one occasion and seized by Prince Kasypa in 1511 when he overthrew his father. The story of Prince Kasypa is fully recorded in legends.

Kasypa was a talented but avaricious younger son of King Datu Sena. He saw no likelihood of succession to power and his father was unwilling to give him a share of his wealth, so he decided to seize both. Biding his time until his brother was in a distant place with his army, Kasypa staged a rebellion. Being a Buddhist he was unwilling to kill his father so he bricked him into a cave to die of starvation. Laden with his father's treasure and the sacred tooth he then hurried southwards with his followers to a new stronghold where he could avoid his brother's vengeance.

He chose well. Sighiri is a fantastic vertical pillar of granite rising 400 feet out of the plain. This massive boss has an acre of ground at its summit on which his fortified palace and terraced gardens were constructed (Pl. 8).

The main rooms of this palace were equipped with fantastic splendour including woven carpets and silk hangings, luxurious couches, gold and silver plates from which to eat, and even bell-pulls with which to summon the servants. Rare perfumes sweetened the air and bejewelled girls gave life to the apartments. By some miraculous feat a throne of red gneiss for Kasypa had been sculptured and hoisted to the centre of his eyrie.

Surrounding the main rooms was an outer corridor from which other rooms and a labyrinth of passages led to the quarters of his staff and his guards.

Outside the buildings a series of flower gardens on terraces led gently downwards to a huge tank for water hewn out of the solid rock.

A memory of the people who lived here has survived in the form of exquisite paintings on the roof of a lofty cave above the western gallery. Noblewomen, with their darker-complexioned servants, are so exactly portrayed as to suggest portraiture. Their bright clothes are decorated with gold chains and jewelled pins, their hair with shining ornaments and their fingers with gem-stone rings (Pl. 9).

The nobles lived at a lower level on a projecting shoulder of rock in rooms overlooking a walled courtyard. The humbler workmen and cultivators had their homes in the plain beside a lake in which artificial islands added to Kasypa's superb view from above.

When standing where often he must have stood with his eyes roving the landscape for signs of an avenging army, I could see solid canopies of forest trees stretching towards the sacred Hill of Mihintale where Buddhism was introduced into Ceylon. Lotus still flourished in the lake below but my hopes of hearing the bark of deer or the scream of a leopard from the sea of jungle beyond were disappointed. Only the monotonous clink of a brain-fever bird rewarded my vigil.

Perhaps Kasypa's principal triumph of imagination and skill lay in the defended stairway up the vertical rock to his palace. Here we need the reconstruction of skilled antiquaries to picture how it used to look in its heyday. A wide sweep of quartzite steps led to an arched gateway situated between the brick and plaster paws of a gigantic lion. Having passed this guardian of the hill, a sloping gallery led upwards where people were saved from vertigo as well as an enemy's arrow by a massive retaining wall of remarkable construction for the 16th century. Being the only route to the summit, this could easily be defended against invaders.

Kasypa had to wait no less than eighteen years before his brother's army arrived and he was then foolish enough to forsake his impregnable fortress to do battle in the plain below. Did the enemy look so small from above, or had time lulled him into a belief in his own omnipotence? In the fighting which followed he lost his life and the sacred tooth was recovered.

CHAPTER FOURTEEN

Krakatoa—An Island is Born

A LITTLE old man with an affliction that bent him double used to shuffle along Cambridge streets after dark with a twinkling lantern in his hand looking just like a gnome straight from fairyland. His destination was St. John's College, where I was to meet him one night when visiting James Wordie and to learn that he was a celebrated volcano expert called Professor T. G. Bonney, F.R.S. Naturally the talk turned to volcanoes and it was from this talk that I gained a burning desire to visit Krakatoa some day.

Lying in the Sunda Strait between Java and Sumatra, this island appeared in textbooks as having caused the biggest explosion in the history of man during which two-thirds of it had vanished into the sea and sky during 1883. The noise had been heard 3,000 miles away and tidal waves had drowned more than 36,000 people within a radius of fifty miles. Ever since this terrible event it had remained as a fascinating example of how nature can restock with life an uninhabited island whose fauna and flora had been utterly obliterated by fire and thick layers of burning cinders and pumice.

By the time an opportunity came to visit Krakatoa some eight years after first hearing about it from Professor Bonney, my ambition was more than achieved. Krakatoa had awakened from its long sleep of forty-five years and in 1928 and had produced a new island called Anak Krakatoa (Child of Krakatoa) from the seabed 600 feet below the surface. I was to be the first person to land on it and any frivolous thoughts of planting a Union Jack were sternly pushed out of mind as being too ill-mannered after all the kindness shown to me by Dr. K.W. Dammerman of Buitenzorg in arranging my journey!

BURMA
THAILAND
VIETNAM
CAMBODIA
Koh Larn
GULF OF SIAM
SOUTH
CHINA
SEA
Penang
MALAYA
Tioman
SINGAPORE
Carimon
SUMATRA
BORNEO
Krakatoa
Sunda Strait
JAVA
N
0
500 Miles
MALAYSIA

The visit lasted from 24th to 28th February 1931 and my companions provided lots of amusement. The alternatives were silence or babel because not one person could converse freely with any other! The crew, for instance, comprised a Malay, a Madoerese, a Javanese and a Sumatran who used different dialects of the Malay language while my two companions were a Russian and a Hungarian. Petroeschevsky, the Russian, provided a choice of Russian or Malay. Liehtig knew Hungarian and a little French. My hopes rested on Liehtig until I found that he was stone deaf from a volcanic bang in Java, so we were forced to exchange written messages in schoolboy French.

Both companions showed me in mime that each thought the other to be a bit crazed and I was inclined to agree, especially with the Hungarian. Petroeschevsky alternated between wild bravado and morbid depression as he gazed at a photograph of the last Czar and reminded himself of the Bolshevik terror from which he had fled into exile. Liehtig smiled sweetly at his inner thoughts about girls or wrote poetry.

Both had had trying experiences during the previous six weeks examining Java's still active Merapi whose scorching gases had instantaneously destroyed an entire village of 1,400 people. Straight from this scene, they started with no joy at the prospect of chaperoning me to Krakatoa. Active volcanoes like Merapi are awesome, but what is feared more by the experts is a Krakatoa which awakes after a long sleep.

Krakatoa lies at the point of intersection of two fissures in the earth's crust along each of which a chain of volcanoes are ranged. Geologists reconstruct its history by saying that Krakatoa, Verlaten and Lang Islands were the remnants of a vast crater wall belonging to a mighty volcano which exploded and collapsed several thousand years ago. This big paroxysm had been followed by gentler eruptions from the submarine crater which had built up the Krakatoa part of the crater rim into a larger island nearly six miles long and four miles wide. These upheavals had been commemorated by three craters on Krakatoa known as Rakata (2,623 ft.), Danan (1,496 ft.) and Perboewatan (399 ft.). The last of these had erupted in 1680 after centuries of silence and since that date the island had once more been completely peaceful until 1883. Despite its fertility it had remained uninhabited and its green hills had become a welcome landmark to the many ships which passed through the Strait between the China Seas and the Indian Ocean.

Some earthquakes had been reported from the Sunda Strait in 1880 but it came as a complete surprise when peaceful little Krakatoa suddenly began erupting in a big way from its Perboewatan crater on 20th May 1883. The earthquakes had caused sea water to seep through fissures they had opened and this, converted into steam, had built up enough pressure to dislodge the rock plugs of Perboewatan. Clouds of steam rose high into the air with explosions vast enough to be heard at a distance of 250 miles.

During the next three months the volcano continued to roar and explode in a terrifying manner. A pillar of steam and smoke flecked with darting tongues of flame rose 3,000 feet into the sky producing a mushroom cloud from which hot cinders and pumice fell on passing ships. By 16th June steam began to rise from a second vent in Danan crater.

The eruptions subsided on 23rd August and all was quiet except for ominous rumblings from the underlying magma as it frothed and bubbled below clogged vents. Pressure mounted during the next three days and finally it gained release with some gigantic blasts starting at 1 p.m. on Sunday, 26th August,

A great bang sent steam and debris high into the sky, dark clouds began to black out the Straits, thunder roared and lightning flashed, ships were covered with hot ashes and were lit by St. Elmo's phosphorescent light in the electric atmosphere, while strange tides rose and fell on the beaches of Java and Sumatra.

A particularly loud explosion at 4.40 a.m. marked the beginning of Krakatoa's collapse into the chasm beneath it and half an hour later a tidal wave reaching thirty-three feet hit the port of Anjer thirty-three miles away while waves of varying height hit other coastal villages. Some were demolished and large numbers of people were drowned. Others had miraculous escapes like one who clambered from a log on to the back of an alligator as both were swept inland!

The explosions and the further collapse of Krakatoa continued until finally half the crater mountain of Rakata slithered sideways

Dotted line shows area of Krakatoa which disappeared in 1883

into the boiling chasm along with quantities of sea water at about 10 a.m. on the 27th August. At this time, 10.02 a.m. precisely, the greatest explosion in man's recorded history took place. Sea water was sucked into the maelstrom and then thrown outwards in the form of a gigantic wave by the cubic mile of earth that had collapsed into the sea.

The sound was heard 3,000 miles across the Indian Ocean,

pressure waves encircled the world seven times, the dust cloud reached seventeen miles into the atmosphere within an hour and was carried three times round the world, and the tidal waves even lapped the shores of America and England at times which suggested a speed of travel in excess of 350 miles per hour.

Near at hand the principal wave wrought terrible destruction and it was reckoned that nearly 37,000 people were drowned along the coasts of Java and Sumatra. The wall of water had risen to between 67 and 100 feet as the shallow shores were approached and rose further to 130 feet at Merak after compression from the two narrowing shores of its bay. People who had taken refuge in stone buildings at the top of a 130 ft. hill behind Merak were all drowned as these buildings were washed away.

At Telok Betong the wave came within six feet of those who had sought safety on a hill of 122 feet. Below them the surging waters had carried a gunboat with a 6 ft. draught more than a mile inland leaving it high and dry at 30 feet above sea-level.

The terrifying experience of those who survived are more difficult to describe than to imagine. They had lost their friends, their homes and everything they possessed. The incessant explosions, culminating in another huge bang as more of Krakatoa collapsed at 10.52, made the ground shake. Inky blackness had blotted out the sun. Thunderstorms raged around them. Tempests blew. More large waves roared on to the beaches. Hot cinders rained and burned them. The rapid rises and falls in pressure made breathing difficult. Many people believed the end of the world had come.

Krakatoa had done its worst and with minor grumblings it now subsided into complete silence for forty-four years. Its long sleep was finally broken in December 1927 by jets of steam and debris, accompanied by explosions coming from a submarine vent 600 feet below sea-level. The explosions became more numerous and violent as January progressed and on 26th a small new island appeared, six feet above the surface which was christened Anak Krakatoa (Child of Krakatoa).

Lulls and renewed activities followed in sequence for many

months during which this island vanished four times below the waves and finally reappeared for the fifth time on 11th August 1930. It was this island, Anak Krakatoa V, which I had set my heart on visiting six months after its birth, but Dr. Dammerman warned me of two possibilities before I left him at Buitenzorg. It might 'softly and suddenly vanish away' like its predecessors—and like Lewis Carroll's Boojum—or the present lull, which was unlikely to last long, might be broken by a fresh series of violent eruptions. These possibilities, Dr. Dammerman said, made it necessary for me to be bound by the decision of my companions as to whether it was too dangerous to land even on Lang Island three miles away. These were chances which had to be taken and we set sail across the thirty miles of sea on 24th February.

Anak Krakatoa being born

Krakatoa's measurements had been reduced in size to about three by four miles during the eruptions of 1883, Verlaten had been increased in length to over four miles by Krakatoa's debris, and Lang Island had remained at less than two miles in length. Anak Krakatoa lay in the submarine crater between these islands, two to three miles from each of them.

A first glimpse of Anak Krakatoa came just as we were passing the white pumice cliffs of Krakatoa and the stern grey crater wall of Rakata where Krakatoa had been split in half. I am not sure what I had expected but a feeling of disappointment was uppermost at seeing what looked like a flat toasted biscuit floating on the sea. The biscuit, nearly three-quarters of a mile in length, was smooth and featureless with an altitude of only 30 feet. However, it was at least still there and no fresh eruption had started, although bubbles were surging up from the sea's depths on its south-western side, showing that the island at this date represented only part of the crater wall.

We landed at Lang Island on a black sandy beach and walked up a dry watercourse where luxuriant ferns and orchids clothed its banks and the foliage of trees formed a canopy of green above our heads. This brought us to a shady platform with a view of the other islands where Petroeschevsky had built a rough observation hut before he had been called away to Merapi. A concrete floor held posts which supported a corrugated iron roof and walls of plaited cane.

Lang Island is practically on the equator and Liehtig went about hatless and naked except for bathing shorts, without suffering any harm, while Petroeschevsky looked as though he was in fancy dress with a wide-brimmed cowboy hat and pyjamas. The intense heat spread into the night so the choice lay between mosquitoes or nearly smothering oneself with a blanket. Bathing from the sandy beach was a joy in such a climate and the shore provided such treasures as the beautiful pearly nautilus shells and the luxury of coconuts which the currents had brought from Java or Sumatra (Pl. 10).

Conversation or other forms of communication were so difficult that the three of us spent much of the spare time with our own thoughts. Petroeschevsky sang wild Russian songs or lapsed into writing notes, while Liehtig read books or smiled sweetly at photographs of Javanese girl friends whose pictures reminded me of a story about an important visitor from Holland. Knowing that he was easily shocked, the local Javanese officials had sent messages ahead of this personage's tour that all girls must cover their bare breasts. The result was disastrous because they obeyed the order implicitly by pulling up their sarongs to cover their breasts.

My companions' tasks were to make observations, write reports and pack up their stores and equipment because the nature of the volcanic debris had convinced them that no danger to Java or Sumatra would arise from eruptions in the near future. It is the plugging of vents with hard siliceous material which usually precedes the really violent explosions and there was no sign of this.

It had been established to my complete satisfaction that the terrible eruptions in 1883 had entirely sterilised Krakatoa of all life, and probably Lang and Verlaten Islands as well since both had been covered with a layer of red hot ashes and pumice to depths of 100 to 200 feet. This made a walk amongst thick jungles in Lang Island all the more exciting, as one wondered how the fauna and flora had managed to arrive across the sea and what was still missing.

During my first day's struggle along dry watercourses and into thick jungle I caught sight of green pigeons in the trees and brilliant little sunbirds hovering round their blossoms; a malay oriole with yellow and black plumage flew across the path and bulbuls flitted from branch to branch above my head. These and the night-time bats had been presented with no serious problem in crossing about thirty miles of sea.

Once I disturbed some rats accidentally imported by visiting fishermen, who would also have been responsible for the house geckoes. Lizards darted out of the way and I did much the same myself when confronted by a handsome six-foot python elegantly posed along a low branch. Frequent rustlings amongst the undergrowth were traced to huge dark-red land crabs which lived in holes all over the island and even up the hillside. A passage through the scrub dislodged quantities of hermit crabs from the branches. The sea was no barrier to them. The pythons and lizards had probably arrived on floating logs and plant seeds could have arrived either by sea or air, but it was strange to find no coconuts growing on any of the islands.

Visits since 1883 had shown a gradual increase in the build-up of the invertebrate fauna so it was fun finding what were likely to be the most recent and unrecorded colonists like a thysanuran, a springtail and an active little bronzy-black tiger beetle. Winged insects and a few kinds of land snails were numerous but earthworms were not found although they had arrived on Krakatoa, presumably in the stems of floating banana or other trees. Wingless though they may be, young spiders of many families can float long distances on their silk parachutes so the significance of my collection of nearly a hundred species lay in the fact that those families which cannot take to the air were unrepresented except for two domestic species accidentally imported by man.

My companions had insisted that the boat could not be used for journeys to Krakatoa and Anak Krakatoa until they had finished their tasks, but when that time came they seemed reluctant to fulfil their promise. So far as I could make out a break in the fine weather was forecast and they wanted to get straight back to Java before a storm broke. The Hungarian smiled apologetically; the Russian waved his arms and raised his voice. In my faulty French I wrote a note for the deaf Hungarian to express my determination either to row across by myself or to remain on Lang Island until the government sent another boat to fetch me. My food could last three days and after that I would have to survive on crabs and coconuts from the beach until I was relieved.

Whether I was bluffing or not, this note did the trick and the day was spent visiting both Krakatoa and Anak Krakatoa.

Krakatoa was covered with even more luxuriant growth than Lang Island. The struggle for light and space was fascinating. It was easy to see how early plant colonists had flourished and then been overtaken and swamped by later arrivals which in turn were fighting a losing battle with more dominant new immigrants. Trees and shrubs grew in one smothering tangle.

Then on to Anak Krakatoa where, with feelings of intense jubilation, I leaped on to its virgin shore composed entirely of dark grey ash, black cinders and white pumice stone. No plants would grow here until weathering and bacteria had had time to create soil in a year or two, but seeds, along with debris like banana stems and other vegetable matter, were awaiting their time to establish themselves. Here was an island crying out for regular study if only it survived. Noël Coward's 'Mad dogs and Englishmen go out in the midday sun' came into my mind as I combed the beach flotsam and raked the cinders for signs of animal

Anak Krakatoa in 1959

life with the sun beating straight down on to my back. Both my companions became infected by my enthusiasm and their ill mood was forgotten. We were all at it together.

The only abundant insects were scavengers which could feed on whatever plant life the sea brought them—a springtail (*Mesira calolepis*) and a beetle (*Anthicus oceanicus*). Ants (*Camponotus* (*Tanaemyrmex*) *variegatus*) might also survive but they were probably stray arrivals like a tiny leaf-mining moth (*Cosmopteryx*) and a mosquito (*Aedes vigilax*). Three predaceous spiders might also survive if insects arrived in plenty or bred in large enough numbers (*Ctenus periculosus, Trochosa reimoseri* and *Maso krakatauensis*).

Thus Anak Krakatoa was receiving both plant and animal offerings most of which it was unready to accept.

At that date the island represented only the northern and north-eastern edge of the crater and the south-western fringe was heated by quantities of gas bubbles coming up from a vent below sea-level. These provided warning of further eruptions before long, but we drew back from the spectacle more because of the hot ground under our feet.

Anxiety about the weather reports proved to be not unfounded. A storm blew up in the night and we got away next day only with extreme difficulty. It was necessary to wade out to the dinghy through waves reaching shoulder height and the journey back to Java was a rough one.

During the next eight years before war broke out, several visits were made by others to Anak Krakatoa and before the end of 1932 fungi, algae and bacteria had been discovered in the soil and twenty species of shore plant had started to grow. Violent eruptions in 1933 destroyed the lot. This pattern of growth and destruction during lulls and eruptions has continued up to the present time. The entire crater rim emerged for the first time in 1933 and gradually the island's altitude had been built up to 330 feet by 1959.

The eruptions and the build up of the island round its crater are likely to continue for hundreds of years before any catastrophe like that of 1883 need be feared. This will be heralded by a change from basaltic to acid siliceous lavas which clog the vents and lead to a period of quiescence. The pressures may then cause earthquakes which will admit sea water and this, converted to steam in a subterranean inferno and mixed with the frothing magma, may then re-enact the scenes of 1883.

CHAPTER FIFTEEN

In Malayan Waters

DURING a passing phase when it is fashionable to denigrate men of importance, I remain loyal to my heroes in the knowledge that if they had never made mistakes they would barely be human. Should anyone find faults in Sir Stamford Raffles these would be heavily outweighed by the merit of his character and deeds.

Certainly he was an empire-builder in a period when that was fashionable and he did nothing but good for the local inhabitants. His vision and courage were always employed to benefit native races besides fostering the interests of his own country. He had seen how the Dutch treated the natives of Java and Sumatra and detested it. He saw how close Holland was to gaining complete monopoly of trade with the Far East and decided to break their stranglehold. His foresight, his actions and his administrative policies were his alone and not those of the East India Company who employed and constantly hounded him.

Raffles saw a necessity to create a base between Malaya and the Dutch East Indies to protect and keep these narrow seas open for trading vessels. A free port between India and China was what he sought to establish. Only a man of vision would have pictured the Singapore Island of his day as the ideal place to select. Eerie mangrove swamps stretching their maze of roots into black mud, and dense malaria-infested jungles, were what he found on his first landing in 1819 to parley with the headman of its small fishing village. His eyes were concentrated on its natural harbour.

Always working with and for the local people, his project was an immediate success, and he was only just in time as the Dutch had already annexed islands like Rhio and Carimon which had been his alternatives to Singapore.

No time was wasted in erecting defences, good buildings, schools, hospitals, courts of law and even a botanical garden. Foresight and courage were combined with a rare depth of understanding, humanity and fairness which brought him warm feelings of affection from Asians of all races who flocked there to share in the island's prosperity.

When independence was granted more than a century and a quarter later, the transfer of government was made from the British to the descendants, be it noted, of those other foreigners who had arrived from elsewhere to reap benefits from the settlement established by Raffles.

My first landing in Singapore was in 1930. By that time the town was vast and still greedily expanding to absorb country residences and villages. The Chinese had outnumbered the Malays by nearly six to one and the island's total population exceeded 400,000. Ships of all sizes and nations lay in the harbour. Fine government buildings, streets of solid offices, hotels, quantities of Chinese and Indian shops all provided an air of prosperity despite the beginning of a world slump. Lean white-garbed Tamils and Telegus, noisy but respectful, looked fitter than in their Indian homeland. Hokkiens and Cantonese from China, always ready for a joke, proudly displayed huge stomachs as evidence of wealth as they grew older. Handsome bearded Sikhs towered over all-comers as they walked with majesty along the streets. The sprinkling of pale faced Britons engaged in government or commerce could be picked out by their white solar topees. The Malays themselves were only visible in the town disguised in the neat uniforms of the police or chauffeurs.

Despite the immense role played by the British in providing sound government and in pioneering Malaya's predominant industries of rubber and tin, it could scarcely be denied that the Chinese formed the backbone of Singapore. Their energies and enterprise were fantastic.

Conversation with the Welsh schoolmaster of a mixed school brought no surprise when he said that a layer of Chinese was at the top of every age group, followed by a layer of Indians with the

Malays in a bottom layer. He attributed this chiefly to physical energy and keenness rather than to innate intelligence.

One piece of his enterprise was amusing. He had translated into Malay a number of English nursery rhymes, and 'Old King Cole' had suitably become 'Old Sultan Cole' who was now calling for his opium pipe, his bowl of curry and his three wives. What more could an eastern potentate desire, except a few more wives perhaps?

Being a Welshman the schoolmaster had gained pleasure from shifting to a Chinaman the stigma of 'Taffy was a Welshman, Taffy was a thief!' Retranslated from Malay his version ran:

> Ah Fi was a Chinaman, Ah Fi was a thief,
> Ah Fi came to my house and stole a bowl of soup,
> I went to Ah Fi's house and found him on his bed,
> So I took a bowl of birds' nest soup and
> poured it on his head!

One soon gets used to a collar sticking limpet-wise to the neck in this humid climate, but for choice the convenience of town residence was exchanged for the subtler charms of the Sea View Hotel several miles beyond its outskirts, where a gentle evening breeze shifted the oppressive blanket of stale air. Dining round a dim-lit dance floor below the stars, with tall slanting palms silhouetted blackly against the sky, one was provided with an atmosphere of delightful romance. And on moonlight nights it was glorious to have a bathe in the square of sea enclosed with wire netting against inquisitive sharks.

Letters to an anxious mother concentrated on early impressions and avoided awkward questions about the use made of the home-knitted cholera belt (which had been jettisoned in the Red Sea before arrival). Recorded impressions before things are taken for granted are sometimes worth while and in these letters the question was considered whether a kidnapped and blindfolded Sherlock Holmes could detect his whereabouts by the sounds and smells of Singapore.

The shrill whistling of the cicadas at night, the creaking of crickets, the toc-toc of the house lizard and the loud croaking of frogs suggested the tropical East. The sharp tapping of bamboos by itinerant salesmen, the clip-clop of wooden sandals and the ear-splitting squeaks which provided companionship to labourers with barrows or bullock carts, identified a Chinese presence.

Turning to smells, there were the subtle blends of earth and plant exudations, the acrid smell of burning mangrove wood, the spices used in cooking and the depressing smell of mould in every cupboard and bookshelf.

The final clue which identified Singapore during several months in the year was a gentle rustling in the trees at precisely the same time every evening followed by the sound of raindrops heralding a sharp short deluge.

The evening attentions of mosquitoes might also be mentioned though these were not nearly so severe as in Bangkok. For the first time I took notice of the maxim about necessity being the mother of invention, which had irritated me so much as a child when confronted with an apparently insoluble childish problem. Discarding local advice that mosquitoes were teetotallers which did not seek blood saturated in whisky, a search was made for remedies more original than the short-lasting efficacy of strong smelling lotions.

An armchair professor had written that mosquitoes were repelled by yellow. Now was the opportunity to test his theory. Diligent search yielded a pair of vivid yellow socks in the market and the evening was spent at the Sea View Hotel wearing one yellow and one black sock with my evening clothes, for in those days all Europeans dressed for dinner (with stiff shirts and collars). Eagerly I awaited the final result and found that a dead heat had been scored on each ankle!

A fresh search in the town produced a pair of thin leather mosquito boots ascending to the calves. These brought immediate respite but were infernally hot.

Long before coming to Malaya it had become obvious that biting and worrying insects vastly preferred me to my friends and suddenly the importance of this fact dawned on me. At least one friend, and I was to meet one other, claimed that in his extreme case he was never bitten by insects. No difference in smell was evident to my senses so where did the preference lie?

Bird species each have their own particular louse which shows no desire to bite any other species, and mammals each have their own kinds of flea. Surely there must be some minor ingredient in the blood, or possibly the sweat, which explains the preference? Millions of pounds are spent in oiling the waters to destroy malarial mosquitoes but this did not affect the other nuisances. How much cheaper and better it would be to discover the cause and inject ourselves with some ingredient which removed the desire of all insects to bite us.

At that time it could be claimed that more people in the world died or were incapacitated from the direct or indirect effects of insect bites than from any other cause, so it was distressing to find that vigorous and repeated attempts to get research organisations interested were all rejected after my return to England in 1932. A brief note on the subject in *Nature* fourteen years later was scoffed at so I finally abandoned my hobby horse.

If the Americans, who have now hit on the same idea, are successful in their researches, who knows what the effects may be? Perhaps the highlanders of Scotland will stop 'flinging' if Sedgwick was right in the one joke occurring in his volumes of stodgy zoology. After speaking of midges which make life unpleasant during a Scottish summer, he allowed himself to remark 'hence the Highland Fling'.

These missions to Malaya in 1930 and 1931 were arduous affairs leading to the formation of a subsidiary company for my employers with offices in Singapore and Penang. Weekends were kept sacrosanct and Owen Eustace was my faithful companion, wherever I decided to explore or go in search of natural history. His knowledge both of Malay and Hokkien languages was a tremendous help.

Even in Singapore the Malays, whose homes we visited, had retained most of their simplicity and knowledge of folklore, but their manners were so courtly that we had to guard against their telling us what they thought we wanted them to say regardless of its truth.

They retained the widespread legend of Mahommedan countries that Mahommed was saved from his pursuers by a spider weaving its web across the cave in which he was concealed. They had however forsaken the method of capturing cicadas employed in northern Malaya. There they seat themselves round a camp fire at night and imitate the male cicadas' love song by tapping thin pieces of bamboo in unison. Amorous females respond, poor dears, by flying down from the trees and allowing themselves to be gathered into the cooking pot. The mild flavour is reminiscent of fresh lettuce.

Growing in Malayan hills are three species of those strange plants (*Nepenthes*) whose pitchers are half filled with a fluid that attracts and consumes insects. A remarkable spider (*Misumenops nepenthicola*) has taken to living just inside these pitchers where it defrauds the plants of some of its victims by grabbing them before they tumble into the fluid. More remarkable still was the discovery that the spider can run down under the fluid with impunity if it is disturbed. A horny skin prevents its being digested and a small fringe of hair beside its lung-books traps a bubble of air to breathe. When the spider emerges after the danger has passed, it simply wipes its mouth against the dry wall of the pitcher and takes up its sentry duty once more.

The dark bathroom at my hotel, with its huge urn of water and scoop, kept me watchful for scorpions but all I found was the fascinating habit of a friendly but slothful spider (*Scytodes*) which made up for its limited mobility by squirting threads of gum from its mouthparts to pinion mosquitoes to the wall.

At low tide one afternoon we visited the reef at Blakan Mati in search of a semi-marine spider called *Desis martensi* which had been recorded without any clues to its mode of life. There it was airing itself on the rocks between tides and spinning a waterproof cell in a crevice before the rising water submerged it. Sandhoppers, a beetle and even a small soft crab fell victim to it while we watched, but if its food was unusual so were its enemies. One strayed in reach of a tiny octopus (*Octopus filamentosus*) and was reaped by a tentacle.

One weekend was devoted to the small island of Carimon in which Raffles had taken a passing interest before he set eyes on the much better base provided by Singapore. A wrinkled old

Chinaman's dilapidated steamer chugged there in four and a half hours and brought us back on the following day with his cargo of dried fish and coconuts. The slump had set in and both the tin mines and rubber estates had been closed leaving a rather forlorn and diminished population.

Carimon had a resident Dutch Consul and a Malay policeman in its only ramshackle village of any size. Both were delighted to have visitors though they thought us quite mad for wanting to climb the highest hill. The Malay policeman had a Dutch wife of massive proportions who declared that not just her feet but the heat was killing her. Having servants and nobody to talk to she spent most of the time fanning herself in the shade of palm trees which lined the beach. To judge by her contours she wore nothing underneath her pyjamas and unwittingly she was setting a fashion for pyjamas long before they were adopted elsewhere as a dress for women. The Consul's domestic bliss never revealed themselves though they disclosed their presence by giggles as they peered at us through a lattice screen.

It was morning when we set out along the coast in the Consul's launch from which we had a good view of crocodiles basking in deceptive slumber on the island's unoccupied beaches. A small group of fishermen's stilted houses was reached in about an hour and here guides were found to lead us to the highest peak known as 'the father of the island'.

The shade of palm trees and rubber estates disguised the true heat of the day until we emerged on the bare ground of a derelict tin mine. Here the scorching sun was thrown back at us by the white soil in all its equatorial strength. Passing beyond this we stumbled in our shorts through a mile of tall razor-edged grasses before a forest was reached, so tangled and dense as to allow passage only by means of a rocky stream bed from the hills.

In this humid twilight we could see a rich variety of orchids covering the mossy branches of the trees. Our route lay upwards and on those occasions when we paused for breath the silence was so extreme we could even hear caterpillars gnawing the leaves above our heads. Then a twig would snap somewhere, birds would begin to twitter and monkeys to swing or chatter overhead.

There is nothing more aggravating than to lose hard-earned altitude through the track turning downwards. Somehow we had lost our stream bed and finally we were halted by an impenetrable wall of scrub. Our guides had lost their way and there was nothing for it but to retrace our steps and start again.

By now we had lost all sense of altitude and direction as we climbed and climbed, streaming from every pore in an atmosphere like a Turkish bath. Suddenly the climb became so steep that it became necessary to pull ourselves upwards by clinging to tree trunks. At last we emerged from the forest and found ourselves within a hundred feet of the bare summit.

Having staggered to the top over loose stony ground we stripped off all our sodden clothes and laid them out to dry on a huge boulder while we rested in the shade. The view over the top of the forest canopy was of an intriguing coastline of distant sandy bays. Over the sea, was the greenery of low-lying Malaya in one direction with the high mountains of Sumatra in the other.

While clothes were drying our guides spent the time collecting long whisks of grey lichen growing on the trees which made, they said, an excellent medicine for expectant mothers.

Our leather shoes slipped on the loose soil on the way down and we had to save ourselves by embracing many a tree. This caused much laughter from our bare-footed guides who chanted 'bus' at every enforced collision, meaning that another form of transport was more appropriate for town dwellers. However they had the grace to laugh at themselves during the final stages of our walk over flat ground when we chanted the same international word as their steps began to falter. Friendship was further cemented with refreshing drinks of coconut milk after arriving at their homes.

Another island which intrigued me was Tioman off the east coast of Johore, but plans for a visit were twice thwarted by fierce monsoon weather and I was left with only an occasional glimpse of its 3,300-foot mountain, the Dragon's Horns, across twenty miles of windswept sea. Friends had described Tioman as an island paradise but this was only hearsay as they had never been there!

Discounting the word 'paradise', I had other reasons for wanting to explore this island of fishermen. In 1890 it had been the last exile of a fantastic adventurer from France called Marie I, King of the Sedangs. His amazing story had been pieced together but his death was still bathed in mystery. Did he really die of snake bite, or had he committed suicide or been murdered? Possibly the oldest inhabitant could provide the final clue even forty-one years after the event. A curiosity had also developed within me to see if the pariah dogs showed signs of descent from Auguste, his poodle, which had gone into exile with him, but my hopes of a visit were again thwarted by weather in 1968.

The forty-eight years of Marie I, King of the Sedangs, had been packed with adventure but it will be appropriate here to deal only with the background to his arrival and death in Malaya.

At the time of his arrival at Singapore in February 1890 he was seen to be a man of remarkable physique and personality. Six feet in height and with broad shoulders, he was obviously a man of great bodily strength. Dark eyebrows above piercing brown eyes and a formidable beard bore out his reputation as a fearless adventurer with a remarkable personality. Stories circulated of his private duels, of valour in the Franco-Prussian War, of skirmishes with hostile tribesmen in unexplored territories and of gigantic swindles which only a man with his personality could have brought off against missionaries, government officers and shrewd businessmen alike.

When draped in the gorgeous uniform of his own design his heroic figure could scarcely fail to impress those who met him. This uniform comprised white trousers and a bright blue tunic decorated with gold belt and fineries at the shoulders. A scarlet silk band passed over one shoulder across his chest to the belt from which a magnificent sabre encrusted with gold, silver and mother of pearl was suspended. The sabre, he claimed, had been a gift from the Sultan of Achim (Pl. 11).

Marie I, alias Marie David de Mayrena, was a dreamer of dreams besides being a man of action. His bluffs were on such a grand scale that all except his victims could find room for admiration or amusement.

His stories of rubber forests, as well as gold, awaiting exploitation in the unexplored independent states of Mois in Annam had hoodwinked M. Constans, Governor General of Cochin China, at a time when the world had developed a thirst for rubber. Finance and letters of introduction to upcountry missionaries were placed at Marie David de Mayrena's service in March 1888 and three months later he emerged from unknown lands as King of the Sedangs with a constitution signed by all the native chieftains! He had suffered every kind of peril and hardship to achieve this end—but had found no rubber or gold.

This triumph brought no joy to France but at first it was hardly taken seriously. Relations between France, Britain, Germany and Siam were extremely delicate in the area East of the Mekong River, so when the Governor General of Cochin China began to experience the arrogance of 'le roi' his uneasiness grew. Marie I King of the Sedangs presented a copy of his signed Constitution, emphasised that his kingdom was an independent state over which no authority from Saigon would be tolerated, offered France its alliance with 10,000 soldiers to march against Annam or Cambodia if needed, and suggested the appointment of a French Consul. The Governor General now firmly repudiated him completely.

Marie's dignified reprisals took the form of renouncing his French nationality, of declaring his marriage to a girl in France as null and void, and of turning against the French missionaries by adopting the Moslem faith.

Relations with France grew worse when it was learned that he was approaching the Emperor of Germany, the British Government and the King of Siam (whom he addressed as' My Cousin') for alliances combined with concessions and mining rights. These approaches met with no response so he then paid fruitless visits to Hong Kong and Tonkin in search of finance before going to Belgium. Here at last his gigantic bluff met with some success by cunningly playing on the vanities of rich citizens who craved for the glory of titles. Ultimate fortunes from the exploitation of his kingdom were dangled in front of them but it was the temptation of immediate glory which brought success in return for money paid into his Treasury. Thus one dignitary became Prime Minister overnight and others Chamberlain, Counsellor of State, Director of Posts, etc.

After arrival at Singapore in February 1890, the King and his Belgian Ministers took up residence at the Adelphi Hotel together with two Egyptian servants, Mohammed and Osman, whom Marie had recruited during the *Sachsen*'s pause at Port Said. Their presence electrified the population of a town where life tended to be monotonous. It awakened chords of romance in some and admiration or half humorous wonder in others, but it produced headaches for the diplomats of several countries.

While telegraph wires were humming, Marie, in search of an heir to his throne, married a Moslem girl called Aisa for whom he claimed descent from a Chiam princess. Her simple black clothes were abandoned for a long tunic of green silk with a scarlet waistband. On her head she wore a white silk turban, with rings suspended from her ears on purple silk.

Although he found no practical supporters amongst the merchants of Singapore he started to recruit Indian and Malay mercenaries, took delivery of armaments for their use and sought a peaceful passage to his kingdom by an overland route through Siam.

The diplomats had all received their instructions by early March and Marie was to learn that Britain could not tolerate the planning of hostile actions against France by residents of Singapore and that they had confiscated his armaments. Worse was to come. He was warned that he must expect immediate arrest if he set foot on either French or Siamese soil.

Singapore had shown that it was tired of him and all roads to his kingdom were blocked. He retired to melancholy musings for a while and convinced himself that before long the British would hand him over to the French for execution. For the first time in his life his nerve cracked and he fled with his Prime Minister and two adventurers to Tioman Island. One was Horace Villeroi, a French sergeant of marines who had fled from Cochin China to escape his debts. The other, an Englishman called Harold Scott who had been discharged from the command of his ship in

Singapore after indiscretions which had caused him to adopt French nationality. Neither had anything to lose and both saw the possibility of gain if they linked themselves to Marie.

For several weeks nobody knew where Marie had gone, but at last news filtered through that he was living on Tioman Island. An agitated headman had crossed to the mainland in search of guidance from his sultan on a matter which was exercising his mind. Never had the headman met such a man as had arrived in Tioman calling himself 'King'. He had the appearance, the headman reported, of a white man, but his hair and eyes were black. He was not English, he denied being Dutch, and his eyes waxed red as he spat on the ground when asked if he was French. He was a wrong kind of white man, a Moslem, having much that was black in him.

The British Resident in Pekan instructed his young District Officer in Rampin to investigate. J. F. Owen was just the kind of Englishman one likes to picture in such a situation. Without delay he sailed the fifty miles from Rampin in his own small boat and walked up towards the King's residence. The house was barricaded and three gleaming gun barrels held him covered through shuttered windows. Calmly he advanced to the door, rapped on it with his walking stick and called out gaily 'Monseigneur le Roi, please have the goodness to open the door'.

Wreathed in smiles by Owen's compliment, Marie unbarred the door. 'You are a brave man,' said Marie in French, 'and courage we admire. Figure to yourself, we expect an army, and see it is a child who comes to us.'

During dinner Owen listened in amused awe to accounts of Marie's past adventures and they became firm friends. Before leaving on the morrow he was able to explain that the government could not allow his Majesty to make Tioman a base of operations against the French, but he had seen enough to realise that this was no longer likely.

Directly Marie's whereabouts became known his Minister of Justice arrived at the British Resident's office in Pekan as an emissary of the other Belgians whom Marie had abandoned in Singapore. They were penniless and wished their petition for funds to be conveyed to Marie. In due course the reply came from the Prime Minister, at 'Le Maison du Roi' in Tioman, saying in courtly language that it was with infinite regret that his Majesty learned of the inconvenience caused to his trusted followers through the tyranny of France and the treachery of Great Britain. He was devastated to hear of their plight but in view of the present financial position of the treasury he found himself unable to comply. However he had much pleasure in acquainting them with his resolution to raise each minister one grade higher in his service.

Snugly tucked away in their agreements was a clause that they would receive nothing until their safe arrival in the Kingdom of the Sedangs, so they sailed home before the mast, sadder and wiser men.

Meanwhile puzzled islanders had been receiving ten franc notes for their wares or services and the supply was running out. Marie was forced to send his Prime Minister to collect money from what was left in the Treasury at a Singapore bank and now it was Marie's turn to be duped. The Prime Minister absconded and the final curtain began to descend.

In mid-September Marie wrote to his young friend Owen saying that Villeroi, an opium addict, was ill and in desperate need of the drug. The letter was followed a few days later by Marie and Scott themselves to report that Villeroi had died and they sought to make an official declaration. Owen made a report, based on their testimony, to the effect that he had died of fever accompanied by stomach pains and difficulties in breathing. Before his death Villeroi was alleged to have dictated a note to Scott which was written down by him and witnessed by Marie and Ah Soon Fok, a Chinese.

Stories that Villeroi had been killed with poisoned fish by Marie, or by Scott, or by both in conspiracy, never developed beyond rumours.

Owen saw Marie and Scott no more until 11th November. Marie seemed to have lost all his bombast and in gentle mood he drew up his will leaving all his property to the Emperor of Germany. Then he went out into the blazing sun with Auguste, the poodle, and his gun. He returned with the news that he had been bitten in the leg by a black snake and within one and a half hours he was dead.

Owen reported that he had found no mark but that he had applied ipeca and ammonia to the spot indicated, besides putting a tourniquet above the knee. Official verdict—snake bite.

Fatal snake bites are rare in Malaya and the bites are always visible to the naked eye, so there is certainly room for doubt as to the cause of his death.

Was this a case of three adventurers falling out and first one, then another being disposed of? Scott was to claim many years later, when safely established in Cochin China, that he had shot Marie in a duel, but the accuracy of such a claim can be discounted in the absence of any gun shot wound. A more likely cause of death would have been poison, but if this had been administered by Scott it would have been sordid murder and he would naturally have preferred to admit to Marie's death being due to wounds incurred in a fair duel.

The conclusion I reach is that Marie I, King of the Sedangs, died of poison but that there is insufficient evidence to determine whether this was self-administered or administered by Scott. My choice would be that he committed suicide when his funds had run out and he saw no ray of hope in any direction.

Visits to Penang came about in the course of my work, but a special excitement for me lay in the opportunity such visits provided to search for an elusive spider whose tremendous importance to scientists had never attracted the interest it deserved. Living plants or animals which are similar to those known in fossil state from times when the earth was young are popularly called 'living fossils' and their importance lies in the evidence of evolution they provide. The discovery of surviving *Metasequoia* trees in China and the Coelacanth lung-fish in the Indian Ocean not many years ago inspired widespread publicity, so why has the Liphistiid spider with segmented body and numerous other primitive features remained with fame unsung? The almost unknown *Liphistius desultor* was closely related to ancestral spiders which had roamed the humid forests of club-mosses and ferns during the Upper Carboniferous epoch 250 million years ago. The buzz of a fly was not to be heard for another 150 million years by which time the old riddle, 'Why did the fly fly; because the spider spied her', had come true. The menace of crawling spiders had been the main factor in the evolution of hopping and flying insects to increase their chances of survival, an escape which in turn led to the later evolution in spiders of silk snares in which to catch their air-borne prey. *Liphistius* was a survivor from the world of crawlers.

Liphistius from Penang was originally described in 1849 from a damaged specimen, and redescribed from another specimen in 1875. None had since come to light. It must surely be there still if only one knew how to look for it and a clue to this had recently been provided by the discovery of a related species in the Malayan hills.

During our first weekend Owen Eustace and I left the town and drove along the coast road to sandy beaches where the warm sea and shady palms provided ideal spots for bathing picnics. The next weekend was spent in the cooler air of the Crag Hotel near the summit of the island's peak. Here we went for walks along a maze of wide paths zigzagging up and down the hill. Special attention was paid to the red and yellow banks of decomposed granite and presently the outlines of a semi-circular trapdoor in a yellow bank was noticed. It was two inches in diameter and unlike that of any known trapdoor spiders in having about nine thick silk lines radiating outwards from the rim like telegraph wires to inform the occupant of approaching insects.

After gently lifting the wafer-thin door and inserting a pipe cleaner down the burrow to avoid losing its course, digging operations began. Down and down we dug with the utmost care until suddenly a huge spider with bright orange-red legs and head leaped out like a jack-in-the-box and lay motionless at our feet. This was *Liphistius desultor*.

Having learned its secret it was an easy matter to discover dozens of these fascinating spiders and to study their habits and structure in detail.

CHAPTER SIXTEEN

To the Barrier Reef

WHEN a tract of land is as immense as Australia, it is better to think of it as a sub-continent than as an island. In this sense Darwin's comment about Australia in 1836 is comprehensible when he said that 'some islands are too great and ambitious for affection'. As a country in which to live I would disagree; as an island I would accept his statement and would turn to the numerous small islands round its coast for enjoyment. Those off the South coast of West Australia include Middle Island, for instance, with its Paper Bark trees half enclosing a pink salt lake as vivid as any cyclamen. But it is the Barrier Reef which outshines them all in colour and natural wonders.

Travelling round the world to Australia and back with the Barrier Reef as a special detour also gave opportunities, for brief glimpses of Fiji, Canton Island, Honolulu and even Greenland in transit.

Greenland, hard, merciless and inhospitable, aroused nothing but a scientific interest in the survival of its people, its animals and plants. Nevertheless it was an interesting experience to stand on its ice-cap desert and to fly hundreds of miles over its smooth unbroken surface which the indomitable Nansen had laboriously crossed for the first time in 1888 on sledges fitted with sails.

Fiji had provided the other extreme of humid heat and luxuriance. Fiji had awakened memories of a job refused in 1923 and of a lecture on it given by Sir Arthur Shipley in my undergraduate days. Sir Arthur, it should be explained, had just suffered defeat at the university senate in his bitter opposition to the recognition of women as members of the university, so he was seen to shudder on finding the two front rows of his audience packed by Newnham and Girton girls.

One of his lantern slides was of an aged Fiji woman who was so grotesquely wrinkled and hideous as to cause titters of amusement. At this he wheeled round to face his audience and said as he stared down at the two front rows: 'You may well laugh, but at least all the women in Fiji manage to get married at an early age.'

After a moment of deathly hush, the girls rose to their feet and stamped out of the room. Just as they reached the door, he added, sadly and almost sotto voce: 'I really do not understand the hurry. The next boat does not sail until Friday.'

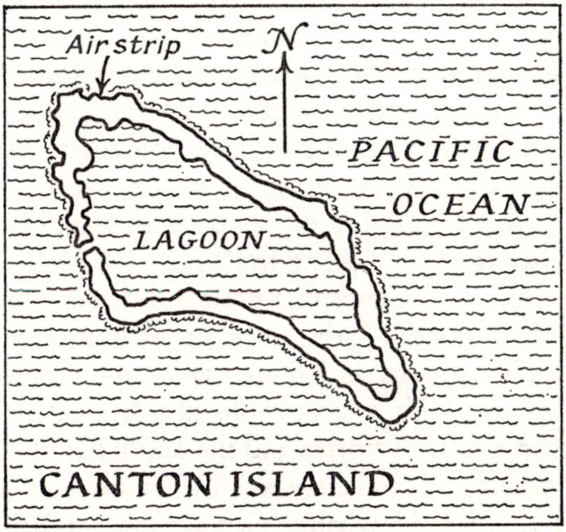

Canton Island was a vision of beauty as we circled overhead in mid-Pacific. Here was a perfect example of a coral atoll built neatly in the form of a closed ring by coral polyps. The bleached wall of coral enclosed an emerald green lagoon contrasting sharply with the deep blue of the sea which encompassed it.

What a disillusionment a landing could bring—a squalid shack where drinks were served; humid heat in which clothes stuck to my body; and that was all. Beauty from a distance had been a mirage.

How different was Honolulu where swarms of loud-voiced visitors had to be removed from one's senses in order to absorb the natural fascinations of the island. Sweet smelling flowers, friendly brown-skinned Hawaiians and of course the unrivalled beaches left memorable impressions of a three days' stay.

Glorious green waves curved everlastingly on their long shoreward journey to break in white foam at their journey's end. Hawaiians balanced on surf boards darted sideways as on skis along the wave crests in order to prolong the period of their exhilaration. Their skill was far beyond the brief visitor's powers to acquire but some compensation could be gained in a Hawaiian-manned outrigger canoe. Waiting for the upsurge of a suitable giant a quarter of a mile out to sea and then the whirl of paddles to remain on the wave crest could not be achieved at the first attempt. In a moment we were capsized and swimming in white turbulence. Success at a later attempt brought the thrill of a sustained onrush at thirty miles an hour with spray stinging one's face and shouts from the boatmen for still greater efforts with the paddles. At last the mighty wave curled over and we shot forward into roaring foam close to the beach.

The string of coral islands composing Australia's Barrier Reef are undoubtedly one of the world's natural wonders. For 1,250 miles off the eastern shoreline countless multitudes of minute coral polyps have built a continuous necklace of islands which serve as a barrier against storms and the South East Trade Winds.

The northernmost island close to New Guinea is called Bristow Island after an intrepid whaler, Captain Abram Bristow, with whom I can claim no kinship. Correspondence between Samuel Enderby and Sir Joseph Banks discloses that his captain was later to discover the Auckland Islands to the South of New Zealand in 1806 and to visit them a year later in a ship with the homely name of *Sarah's Bosom*.

The coral islands of the Barrier Reef are submerged by the sea at high tide but between them and the mainland is a chain of granite islands which owe their origin to a sinking of the coast level. It was to one of these called Hayman Island in the Whitsunday group that we set out with Jeff and Hazel Ruttledge of

Brisbane. Here we occupied adjoining bungalows beside a beach of dazzling white sand and went for meals to the hotel's communal dining-room. Behind the hotel were pine trees and dense thickets of *Pisonia* whose smooth grey trunks have canopies of sage green foliage. The thicket was edged with strange *Pandanus* trees weirdly supported by a maze of stilt-like roots, their stiff clumps of lanceolate leaves contrasting strangely with the delicate foliage of *Casuarina* trees.

The day was hot and unthinkingly we plunged into the sparkling sea. Here we were hotly pursued by our hosts whose feelings of anxious responsibility made it necessary to complete the bathe in three feet of water. Sharks were, of course, their special fear but with some interest we were later to see a couple of sting-rays flapping their ungainly black bodies in shallow water near the landing stage.

After delays caused by weather we eventually set out by launch on a falling tide to visit the reef. The first glimpse of coral gardens was provided by gazing downwards through crystal-clear water at exquisite shapes, dazzling colours and quantities of bright fish. Presently as the sea got shallower the details could be picked out more clearly. Brown coral and waving seaweeds formed a background to other corals of light blue, lavender, yellow, green, pink and rose red. Some had grown as neat cushions, others as patinated mushrooms rising in tiers, or delicate fans, or irregular stags' horns. Some had green branches tipped with pink buds and others yellow stems bearing blue tips like exquisite flowers. Besides these hard corals were rubbery convoluted layers of soft alcyonarians swaying gently in the current.

Life teemed in grottoes formed by the coral and enhanced their beauty. Brilliant sea anemones, large blue starfish, olive green brittle stars, dark purplish sea urchins and even the worms with long scarlet tentacles projecting from their sandy casements had become beautiful in this Garden of Eden. Large and small fish patrolled the gardens or dashed to cover in dim mysterious caverns. Some were sober in colour but others were scarlet, green or blue with golden tails, vivid stripes or spots.

After scrambling on to a small coral island, we moved from pool to pool probing and prying on the inhabitants. Here the clams were small with bright green or blue mantles between their convoluted shells, but those in deep water grew to a quarter of a ton in weight and became the largest shell fish in the world.

Glazed brown-spotted cowries up to four inches in length were concealed under rocks in the deeper pools and the discarded opercula (or lids) of the turban shells seemed to gaze up at us like eyes with green pupils. These 'Cats Eyes' are treasured for local jewellery.

Having once already caused anxiety to our hosts, we listened attentively to all their 'do's' and 'don'ts', and promised to be good. There may be no apples (or apricots) in this Garden of Eden but there are numerous other hazards.

The clams which can close like a vice on the thoughtless finger or ankle of a sluggard. The rare but fatal stings of a tiny jelly fish and of a small octopus. The prickly fish with thirteen deadly spines on its neck. The beautiful cone shells which can cause paralysis with a knife-like tooth that strikes into the palm of an unwary collector's hand. Fortunately all the beauties of these coral gardens can be enjoyed without risking these dangers.

A warning against one small hazard, however, was not heeded with success. A leg and one thumb were slightly scratched by coral and the minute spicules caused discomfort for a week and would have festered if left untreated. Were the ship-wrecked mariners of Ballantyne and Defoe so tough as to take no notice of the cruel sores they must have sustained when cast on coral strands, I wondered?

These hazards, other than the accidental effect of coral scratches, are essentially protective devices and not designed specially for man's discomfort. Fierce competition is waged between the teeming inmates of the coral reef and some means of protection from enemies are extremely bizarre. Teasing a number of sea anemones with a pencil eventually gave the result which was hoped for. A flash of silver marked the exodus of a tiny fish from its remarkable haven inside the anemone's folds. Another tiny green fish is known to occupy a rude position inside the hind gut of the sea cucumber, but the only response to teasing was the equally rude and remarkable response for which bêche-de-mer are famous. It ejected its stomach, an action which happily precedes the growth of a new one!

Spider crabs, skilled in the art of make-up with pieces of clipped sea weed as camouflage, sidled across the deeper pools, and a tiny pea-crab was found in its customary hide-out inside a living clam's shell.

With the tide creeping up on our island we reluctantly tore ourselves away from this amazing display of glamour but during the homeward journey we were reminded once more of the cruel competition which dwells amidst such beauty. Flying fish were escaping from unseen marauders and frigate birds above our heads were forcing the gulls to disgorge their fish which they then caught before they fell into the hungry sea.

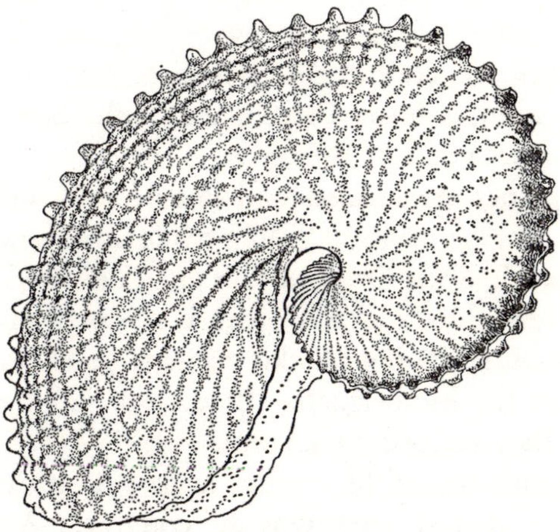

The Paper Argonaut

CHAPTER SEVENTEEN

A Labyrinth of Islands

PLACES can change for better or for worse, but impressions of Manila in 1961 were of a large town sadly lacking in character or integrity whose veneer of civilisation thinly disguised much that was primitive and degenerate. To the visitor it was one of the most expensive and to the resident among the most corrupt places in the world.

Corruption at the top had its miserable echo of rackets and lawlessness which brought us close to danger on more than one occasion during our month's stay in the Philippines.

The object of the visit was to use the principal island of Luzon as a base from which to explore a few of the 7,000 islands of which the Philippines are comprised. These scatter into the Pacific Ocean, the South China, Celebes and Sulu Seas, and stretch almost to Formosa and to Borneo.

The Philippines are blessed with fertile soils and seas so the population is amply provided with its staple diet of rice, vegetables, fruit and fish. Thickly-forested hills and cultivated plains are to be seen in every direction and green is the predominant colour with cogan grass springing up like magic to patch any temporary scar caused by the lumbermen. Beware of poking your finger into the soil, they say, lest it take root.

A facile reminder of Philippine history is provided by the headgear of the men in the towns and country. A scarf knotted over the head indicates Malaysian ancestry. Pagoda-shaped hats of plaited palm, often worked into exquisite patterns, can be attributed to associations with the Chinese since ancient times. Wide-brimmed straw hats are reminders of Spanish domination from 1581 to 1898. Ugly peaked skull caps mark the period of American control from 1898 to 1935. The Japanese wartime

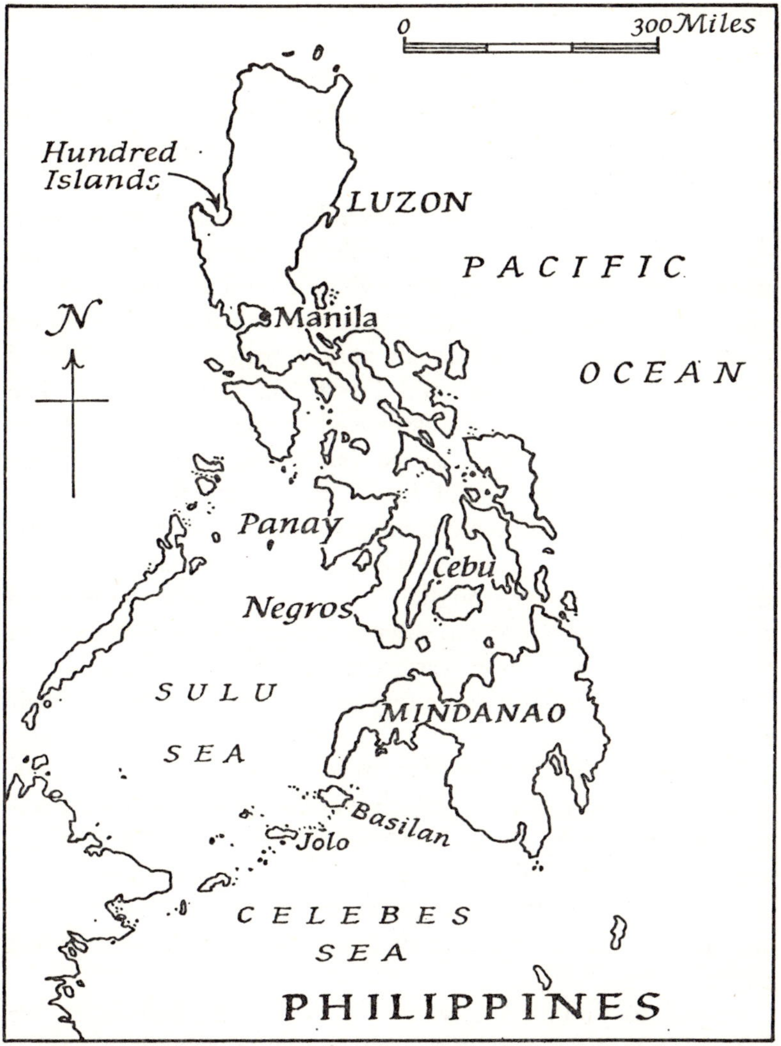

occupation between 1942 and 1945 had left no mark except for a lingering hatred.

One quest in the island of Luzon was for a cave previously unexplored by naturalists or archaeologists and the choice arrived at after consultation with the Director of the National Museum produced experiences quite different from the ones we had expected.

Helen and I set out by jeep in the company of two guides from the museum, one of whom was armed with a shot gun on account of the bandits who were said to be roaming the hills thirty miles outside Manila.

After passing through much splendid mountain scenery with deep valleys and thickly-forested hills, the jeep dived down a side track to the bottom of a valley where a clear mountain stream gushed past a small group of attap houses. Leaving the jeep we now walked in single file along a narrow path where swallowtail butterflies sipped at the wild flowers in the sunlight. The peaceful scene was suddenly destroyed by the appearance of two fierce-looking men who barred our further progress. They clearly had no intention of letting us go any further and an argument developed between them and our two guides, accompanied by menacing gestures on both sides. The path, they said, led only to the cave we sought. The cave belonged to them and they would allow nobody to enter it, not even bat-collectors from the museum. No reason was given, only a blunt refusal.

Our guides had become angry at this loss of face and their resort to menaces had been met with threats of violence in reply.

Conciliation and an offer of money met with no greater success, so a deadlock was reached which was finally broken by a compromise offered by these guardians of the cave. They would send us to an unexplored cave of which they had heard from aboriginal negrito tribesmen, just round the other side of the hill, and they would provide a guide from one of the houses we had passed.

Suspicious though we were of their strange behaviour, we accepted the offer and were soon marching behind this guide through a tapioca plantation with our own armed guide in the rear in case of trouble. An animal track led into the humid gloom of the forest for about three miles before the guide suddenly left it to clamber up a steep rise to the base of a limestone crag. At first he could find nothing, and admitted he had never been there before, but eventually, he spotted a narrow crevice at ground level not much more than two and a half feet in height. Although this looked far from promising, we followed his example and rolled sideways into the inky blackness beyond. The squeaking of countless bats above our heads proclaimed that a lofty chamber had been reached, so we stood up and shook our clothes free of bat guano and cockroaches.

Powerful torches were now switched on and the sight revealed by their beams will never be forgotten. There, in front of us and on both sides stretching far into the distant shadows, lay or sat rows of human skeletons in various postures. Withered flesh and shreds of clothing still clung to them and some were partly buried in bat guano.

This was a moment of horror. We stayed only long enough to identify the bodies as Japanese soldiers from the steel helmets strewn beside them before regaining the freshness of what had previously seemed to be the humid forest gloom.

The guide seemed to be thoroughly shaken by the discovery and he now surprised us by talking in English and trying to make amends by telling all he knew.

This area had been one of the last strongholds of the Japanese in 1945 and many had lingered on in bands either because they did not believe the war was over or in fear of disgrace. The negrito tribesmen had reported that many Japanese had died in the forest from wounds, disease or their own hands and we had obviously found their last hospital or mausoleum.

Our guide and his friends were all from Manila with no interest in the neighbourhood except for the cave we had intended to visit. This they guarded night and day while they dug for loot supposed to be buried there by a Japanese General. This General had arrived with vast treasure from Indonesian temples and had made that cave his headquarters so it was believed that he had buried his treasure there before dying.

The story of the General's treasure from Indonesian temples was confirmed by the Director of the National Museum after our return to Manila, but it has not yet been found and is likely to remain as a local legend for centuries.

The time had now arrived for a cruise to the islands of the South and we boarded the s.s. *Elcano* in time to sail at midnight. She was to dodge in and out of four seas taking supplies to Iloilo, Negros, Cebu, Mindanao and Jolo, and collecting their sugar, copra, timber, hemp and other products for the Manila markets. There were reasonably comfortable cabins for eight passengers, each equipped with a shower, but they were swarming with cockroaches and Helen was unfortunate enough to contract dysentery during the voyage. Almost unlimited passengers slept and fed on the lower deck beside their caribou, pigs and chickens.

We were wakened by the crowing of cocks to see an inquisitive caribou leisurely licking our forward porthole and gazing in at us with a vacant expression of surprise. The view from the deck was of an indigo sea stubbed with large and small green islands. A shoal of tuna were leaping and splashing near the ship, Moros in their flimsy outriggers were busy fishing far from land and the fin of a shark occasionally broke the surface of the sea. Not a single bird was in sight due to the constant egg-collecting and trapping by the Moros, and throughout the cruise we were only to see an occasional sea eagle and one or two solitary gulls and terns.

Many of the Filipino deck passengers were being settled in Mindanao so they had all their possessions with them including their prized fighting cocks. One was nursed under each man's arm and no doubt these would be engaged in deadly combat during the national Sunday sport of cock-fighting.

Well-behaved caribou were in a separate enclosure. 'How they smell,' said one of the passengers. 'Yes, of money', replied the captain, whose own appearance strangely resembled that of a good-looking bull.

Caribou largely take the place of cows in these islands and during a game of chess the captain spoke of a legend reminiscent of a fable from Aesop. The cows and the caribou went bathing together one night. No moon was shining when they came out, so in the darkness they made the mistake of donning each others' skins. This explains why the cow has had a loose skin ever since and the caribou a tight sleek one.

Later, as we were approaching the harbour at Panay in the island of Iloilo the captain showed his knowledge of local lore by telling us about seven tiny green islets just outside it. These, he explained, had sprung up where seven eloping maidens had been drowned in their lovers' arms and they were called the Seven Sisters. Their smooth surfaces resembled funereal mounds and on one was a small white lighthouse like a monument erected in their memory.

Rounding the large island of Mindanao, and calling at four of its harbours with the attractive names of Davao, Cottabato, Dadiangar and Zamboanga, we floated over 35,440 feet of water at the famous Philippine Deep whence we could also see Mt. Apo, the highest mountain, with an altitude of 10,827 feet. After loading Manila hemp at Zamboanga the ship passed into the Sulu Sea through a narrow strait beside the island of Basilan.

Basilan's green hills looked peaceful enough so it was strange to learn that its lawless inhabitants had caused a ban to be placed on the ship putting into its harbour. A regiment had recently been stationed there to suppress banditry, but the latest news was to the effect that the soldiers had found it safer and more profitable to remain within their barrack walls.

A course was now directed to Jolo which had an even more evil reputation. Jolo lies in the Sulu Sea close to Borneo and it had had piracy in its veins long before Rajah Brooke's time. Successive rulers of the Philippines had never been able to tame its inhabitants. They had remained loyal to its Moslem faith while over eighty-five per cent of the Filipino population elsewhere had been converted to Christianity. It had seemed sensible to the Americans during their period of rule to elect the Sultan to Congress in Luzon, but the effects had been the opposite of what they had sought. The removal of his local influence had reduced the island to a state of complete lawlessness. Peaceful farmers went in fear of robbery or death, smugglers were attacked by pirates and pirates by bandits or outlaws.

The introduction of a substantial police force restored some semblance of order in the port itself during the hours of daylight, but their influence had ended there. It was said that only one man could travel in safety more than two miles outside the harbour area and he was the local doctor, a man of mixed blood with a German ancestry. His son, Frederico Kiel, happened to be a passenger in the *Elcano* and he offered to show Helen and me as much of the island as he dared.

Considerable activity, both on board and on the quayside, preceded *Elcano*'s arrival. The ship's bar was not only closed but swivel stool tops and other movable objects were all stowed away. Cabins were locked, fire hoses were brought on deck and officers had revolver holsters strapped to their belts.

On shore we could see barricades being erected, an armoured car standing in readiness for trouble and police in steel helmets carrying rifles.

Beyond the barricades were an open market with stalls, a rabbit warren of dingy houses and one dusty road leading through flat cultivated fields towards humpy crater-like hills about two miles away. To the left of the bay was a fringe of the usual coconut palms and banana trees marking a Moro fishing village of stilted houses whose inhabitants did most of the smuggling of goods from Borneo.

Directly *Elcano* was tied up alongside, the Chief of Police came aboard to post half a dozen armed men at vantage points on the decks. His second task was the delivery of parcels of contraband to the ship's officers for smuggling to his friends and relations in Manila. After that he had time to recognise Frederico and to be brought over for a talk with us.

His job, he emphasised, was to keep order in the town and we could shop safely in the market where we would find imported cigarettes of every brand, watches, scent, radio sets, textiles and much else at cheaper prices than in England or America. There was no sense, he said, in reducing everybody's income in Jolo or raising the cost of living by trying to interfere with the traditions of smuggling which had always prevailed in the island.

When he heard of my keenness to make some examination of the natural history outside the town he told us that on no account must we travel more than two miles outside it. If we did this and did not return he would not even send out a search party to look for us. Wear no jewellery, he warned, and a still greater incentive to attack would be a revolver.

Dr. Kiel was out on his rounds so Frederico accompanied us to the Moro village of smugglers. The stilted houses were built out into the sea and reached by bamboo bridges which could be removed like drawbridges in case of trouble. The people were of dark Malaysian stock and seemed friendly enough as they paused from their domestic chores to stare at us. Some were cooking or mending fishing nets and others brewing spirit from palm milk, while naked children splashed about in the shallow sea.

Frederico pointed out one fierce-looking old man to whom his father was periodically summoned for sedatives when he went berserk. On such occasions he sought a Christian to kill. Today he looked peaceful, but, whenever the mood came on, his friends tied him up with rattan cords in the knowledge that a berserk Moro in search of blood would be shot by the police like a dog with rabies.

Frederico next drove us through fields of sugar cane, hemp, vegetables and fruit to the base of the humpy hills we had seen from the ship. This marked the two mile limit of safety so he turned the car but allowed me to get out to prospect the scene. A dusty track wound up the hill slope amongst the trees and I could see butterflies flitting in the gleams of sunlight. I took my net out and went in pursuit. Nobody was in sight but I had scarcely gone a hundred yards into Tom Tiddler's Ground when five armed men appeared like magic beside me. Each had a long knife and a mother-of-pearl-handled kris. Not having anything of value on me I decided to stand my ground and concentrate on the butterflies while Frederico started the engine and backed a little closer towards me. The guardians of the hill seemed amazed at my antics and examined my net and its captives with the utmost curiosity. Frederico and Helen joined me, and with expressions of goodwill we managed to disengage ourselves with honour satisfied.

The novelty of a butterfly net was thought by the Chief of Police to have been a far better passport than either a revolver or a white flag!

Having returned safely from bandit country we came into the den of cut-throats thronging the market place. Here hard unsmiling stares assessed our worth under the wary eyes of armed police. Never before or since have I felt encompassed by so much evil as I spent the small amount of money in my pockets on known brands of English cigarettes at only a fraction of the price at home.

Groups of fearless children encircled us at the outskirts of the market and provided light relief with their fearless questions and reactions to our answers. Were we Americans? How rich were we? What was our religion? (Sighs when we said 'Christians'.) Did we have coconuts in England? (Groans.) How many sons had we got? (More groans and a chorus of 'bad luck' when we said 'three daughters'.)

The atmosphere of menace in Jolo had been an interesting experience, but it was as much of a relief to the captain as to the passengers when the ship sailed before darkness set in.

A few days were spent in the hill town of Baguio after returning to Luzon and then we drove down to Lucap and Hundred Islands on the west coast. A hot dusty road ended in a concrete square bounded by a sea wall over which the slanting masts of outriggers could be seen at their moorings. To the right stood an attractive white-plastered restaurant with green-trellised windows, whose annexe with eight bedrooms stood on stilts over the sea beyond it. This had a nipah palm roof, a wooden frame, walls of plaited bamboo and springy bamboo floors.

Each room had unexpectedly comfortable cane beds covered with a thin kapok mattress. The water from the shower sped freely through the bamboo flooring whose chinks also provided glints of the sea and sometimes the dark eyes of a child pausing from crab collecting to indulge its curiosity.

Mrs. Lichauco managed the hotel with the help of several daughters and, when not engaged by domestic chores, they prepared a medicine called aro sip from a seaweed collected from the rocks at waist depth. This weed was mixed with oil and vinegar before being sold in jars to the many who suffer from goitre in the hill districts.

When evening came the Lichauco family settled down to play their national game of Songa. Two people share a board with two rows of slots and a larger slot at each end. According to certain rules, bleached cowries are moved with lightning speed from slot to slot with the object of accumulating more in the end slot than the opponent while still retaining some in the small slots in preparation for the next game. To the unskilled observer the game seems interminable but battle with an expert soon revealed the skill.

Beyond the semi-circular bay, in which patterns of posts marked a complicated system of fish traps, protection was provided by an archipelago of low green islands merging at a distance into one humpy line resembling a smooth-crested sea serpent. These were known as Hundred Islands.

The sea serpent dispersed into innumerable tortoise-backed islands varying in length from a few yards to a fraction of a mile as we scattered the flying fish in a hired outrigger. The original outrigger was a hollowed tree trunk with lateral bamboo poles in the water to keep it from rolling over. Ours was what might be called a Mark III with a lightly constructed narrow hull, a slanting mast and an outboard motor, with an outrigger contrivance of three thick bamboos lashed over each side to stout poles in the water.

Within half an hour we were twisting and turning amongst this multitude of islands, sometimes squeezing through channels so narrow that a biscuit could have been thrown on to two islands. All had green tops and the larger ones lured us with coral sand beaches with beautiful sea shells.

One remarkable feature was the overhang of the rock fringes. Sometimes we would see an island with a long roofed slit through its middle at sea level only to find on closer approach that there were really two islands each with an overhanging ledge stretching out several yards above the sea. The effect was extraordinary and often gave a tiny islet the shape of a mushroom with a convex platform supported on a narrow stalk of rock.

I had a childish urge to sit on one of these mushrooms like the caterpillar in *Alice in Wonderland*. This meant wading the last few yards and then finding the task far from easy. The edge of the mushroom's cap was at shoulder height and obviously there were no footholds owing to the overhang. Arms reached over the edge found extremely uncomfortable finger holds in sharp-edged cellular rock. The cutting edges of the rock bit viciously into the knees and legs as they were laboriously heaved on to the cap. Even the smallest islet had prickly shrubs and the glassy edges of the rock cells made any seat like a bed of nails (Pl. 12).

Three happy days were spent in lonely isolation exploring, landing, bathing and collecting beautiful sea shells, but in spite of their great beauty a pre-disposition to love coral islands in sun-drenched seas leaves a certain feeling of disillusionment. Such islands are lovely to look at from sea or air, and some retain their beauty after landing, but they leave an impression of unyielding hardness. The parched ground has no spring in it. The sand is harsh and the vegetation is prickly with none of the gentle wild flowers of temperate regions. Lurking dangers on land and in the sea limit complete relaxation.

Such islands remind me of some beautiful woman on whom it is a pleasure to gaze but whose heart is made of stone.

CHAPTER EIGHTEEN

Of Danger in the Gulf of Siam

THE discovery that more serious hazards are encountered in leisure hours than during a full-time job in an explosive factory often has its parallel when travelling in the wilder parts of the world. Bandits, hostile tribesmen, poisonous snakes and disease carrying mosquitoes or tse-tse flies can all be imagined or met, but fear and greater danger is more often manifest when careering in a ramshackle car along a narrow track beside a precipice. Real danger sometimes descends when least expected and is often quite mundane.

An illustration is provided by travels in Thailand with my daughter Ginie in 1964. These had included a journey by boat from Chiengsen down the Mekong river through rapids with trigger-happy communists on the further bank. She had also been the second white woman ever to meet the almost legendary hill tribe known as the Spirits of the Yellow Leaf during an expedition into the jungles of Nan. Back in the south again we were spending a peaceful weekend at Phattya on the shores of the Gulf of Siam when it was decided to spend the day on a small uninhabited island called Koh Larn visible out to sea.

Two young Thais were hired in their tiny Japanese speed boat who, to our great surprise, were also able to supply Ginie with a pair of water skis. With scarcely a ripple on the sea we sped outwards with Ginie on tow, though a less confident father might have had qualms about sharks in the event of any mishap.

Having landed on the further side of Koh Larn in a bay of white coral sand after a journey of twelve miles, it was arranged that the boatmen would return at four o'clock.

On close inspection the island's green covering dissolved into prickly scrub growing in parched red soil. A climb to the highest point provided no feature of interest so we returned to the seclusion of the sandy bay which was shared by a few black crows.

The sea was of a lovely transparent blue except for a mauve band some sixty yards from the shore which obviously marked a reef of coral. The sun was intensely hot and the water deliciously warm as we swam out to inspect this with snorkels. Plenty of inconspicuous brown fish were seen on the way but directly the coral was reached the scene changed abruptly. Here the fish and other denizens took all the vivid shades of green, blue, pink and red displayed by the coral itself, as though by some photographic process.

The sight was not enjoyed for long because discretion hastened our return to land when a shoal of silvery fish began leaping out of the water in our direction, warning us of an approaching shark or other predatory monster. It is remarkable how vigorously one can swim with a suspected shark just behind one!

The white coral sand was strewn with cowries, pink mussels and other attractive shells but none of the irresistable giants of the tropical seas, so we settled down for lunch and a drowsy afternoon of basking and bathing in the shallow water until the boatmen returned promptly at four o'clock.

Both the boatmen, it seemed, had been bathing with their clothes on, but the only thing they could convey to us by signs was that they did not want Ginie to return on water skis. The reasons for this and for their soaked condition were immediately apparent when the boat rounded the headland. The contrast was fantastic! Here a fierce wind blew into our faces and piled up large waves which crashed against our cockleshell's side until it had cleared outlying rocks and its course could be changed to face them. Now we bobbed up and down like a cork over their crests and down into the valleys beyond making slow progress into the open sea.

Once we were free of all shelter the waves increased in size and the wind brought continuous spray into our faces. Within minutes we were soaked to the skin and holding on to avoid being jolted too violently or thrown overboard.

It was fascinating and reassuring to watch the boatmen coping with the oncoming waves which rose like high walls and swept towards us. Down in a deep trough there was nothing to be seen except the next wave. Acceleration, and a last moment curve sideways to avoid head-on collision took the boat up to the wave's crest and momentarily along it, before changing course and plunging downwards into the deep trough between it and the next wave. One slight misjudgement in angle brought a sheet of water right over us. Many more like that, we knew, would sink us, capsize us or stop the engine, and at this distance nobody on the sparsely populated shore five miles away would notice.

There was something both enjoyable and exhilarating about our plight as we soared and dived over the waves with the cool spray biting into our cheeks and our eyelids lowered to avoid a bruising of the eyes. Water poured down our faces as we frantically bailed with two empty tins.

Whoosh! And all our handiwork was undone as another wave swept over us. But that was to be the last before the boat came safely to shore.

CHAPTER NINETEEN

My Favourite Islands belong to Wales

THE infernal cheek of my first visit to Skomer in 1929 still fills me with shame. Having hired a boatman to take me across from Marloes, I had marooned myself for three days with the intention of spending that time in the only house with Sturt and his family whom I had never met. It was true that a letter had been written asking if he could have me as a paying guest for the Whitsun weekend but the absence of a reply should have been taken as an ominous hint.

A walk up the hill slope amongst late primroses and sea pinks brought me in sight of his house across a saucer of land filled with a glorious sea of bluebells and a small lake tufted with yellow flags. This was obviously the kind of island where Katherine Mansfield would have said: 'I could weep like a child because there are so many flowers, and my lap is so small, and all must be carried home,' but at that moment my own delight was cut short by seeing Sturt watching me advance with suitcase in hand. Feeling as sheepish as I deserved, I was asked who I was, whether I had not received his letter saying a stay would be inconvenient, and whether that was my boat disappearing into the distance.

Having grudgingly accepted the inevitable, he and his family could not have been kinder and I was to return a year later at their own invitation.

Sturt was a lean man of about fifty with a high-bridged nose, a moustache and a most remarkable front tooth from which it was difficult to deflect one's gaze. The tip bent outwards at right angles. This was specially difficult to explain in a retired dentist from Exeter, unless it was a skilfully-invented fabrication for the gripping of his pipe stem.

Sturt had certainly not chosen this lonely life for the sake of his wife, who was frail and gentle, nor for his fourteen-year-old daughter, yet it was difficult to find reasons for his own choice. He was neither handyman nor naturalist, fisherman nor cultivator. Weeds competed freely with vegetables in his garden and work connected with his few livestock was a bore. His conversation was of the town, of people and of cricket. He had decided, it seemed, to forsake a profession he disliked and had calculated that life on Skomer would be cheap. Although he was no artist, he was conscious of beauty; although no naturalist, he liked to see the birds around him; and then there was the intangible feeling of being monarch of all he surveyed.

Solitude was not so likely to produce the same state of tranquillity in a growing girl and at a tender age she was to elope with a neighbouring lighthouseman with whom, it is hoped, she lived happily ever afterwards.

With an area of 750 acres and a length of about two miles, Skomer is the largest of the group comprising Ramsey to the north and Skokholm to the south apart from some tiny satellites further out to sea.

The first walk opened my eyes to the variety of scenery provided on so small an island.

Bluebells are usually associated with woodlands yet here they throve on sheltered slopes beside the solid mass on the open plateau. Nor were they ordinary bluebells with their dozen to twenty bells. Sturt told me that the record topped the century and I myself was to find one with ninety-six bells on a single stem.

From bluebells we came to short turf gleaming with pansies, violets and eyebright. Wheatears flitted ahead of us and larks sang their hearts out in the May sunshine. Black, fawn and an occasional white rabbit stared at us before drumming the warning of our approach to their fellows. Clumps of golden gorse and low heather not yet in flower were passed before reaching the exposed slopes on the western side. Here the massed cushions of sea pinks provided comfortable seats from which to admire the wild cliff scenery where guillemots, razorbills and gulls were joined by jackdaws and buzzards.

Descending to a cave, we stood in twilight with backs pressed uneasily against the wall while huge Atlantic seals came slithering and flapping past without seeming to notice our presence. Perhaps our entrance had disturbed the suckling of their young, but without a torch we were unable to explore.

Circling back to the more sheltered northern frontiers, it was noticeable that the hard igneous rocks of the island's core gave way to quartzites, conglomerates and breccia together with quantities of the island's special basalt called skomerite. Samples of this were added to the island's other speciality, the Skomer vole, but being the kind of naturalist who prefers seeing to collecting, I soon released this inoffensive little animal. The rest of the vertebrate fauna, other than birds and two kinds of seal, was found to be house mice, shrews, lizards, slow worms, frogs and toads.

A second visit to Skomer in June of the following year was shortened by a day when boatmen, over-timid as I thought, refused to cross the short expanse of water on account of tides and winds. Later experience was to teach me that they knew best and that even the islanders living on Skomer, Skokholm and Ramsey had all had their boats smashed or narrowly escaped being swept past the islands into the open sea. The tide races are at their worst near to Ramsey's only harbourage.

Ramsey, originally Raven's Isle, is reached from St. David's where the boatmen provided the local gossip about the Lewis family who had agreed to accommodate my cousin Bobbie and me. Our boatman obviously admired their courage more than their farming skill and said their struggles were unlikely to continue much longer. Besides rats eating their corn and rabbits their pastures, the fates had struck several bitter blows. Lightning had killed some of their horses and cattle, cows and sheep had fallen over the steep cliffs, no less than three boats had in turn been smashed, and now a bull had suddenly gored one of the sons.

A distress signal had been lighted on the hill to summon help and old Mrs. Lewis was still on the mainland with her daughter and injured son, leaving the other son to fend for himself. This, the boatman added with a wry smile, might relieve us of embarrassment, because old Mrs. Lewis had developed a fixation that each visitor would turn out to be the Prince of Wales arriving at last to fetch her daughter away.

Ramsey's twin hills range up to 400 feet and it was in every way a grander and more interesting island than Skomer despite its

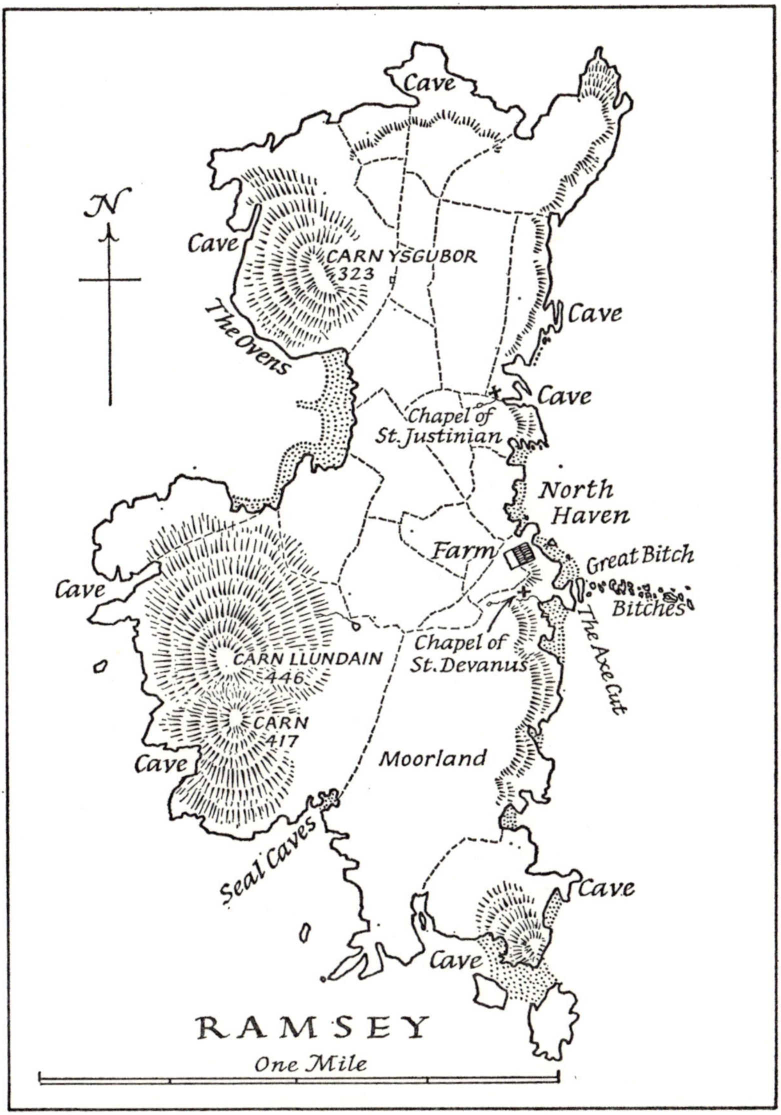

smaller size. The hills were higher, the cliffs more magnificent, the bluebells even more glorious, the history better documented and the tide races more awe-inspiring. A narrow wedge of rock called the Great Bitch was separated from the island by a narrow channel just below the Lewis's house, and its suction during the rising tide had drawn many a small boat to destruction. Water rushed through the bottle neck with terrific force before dividing into two races, one of which passed through a natural rock arch. It was a fierce sight and the roar of its churning waters could be heard in the house twice in every twenty-four hours.

Half a dozen buzzards were soaring overhead as we climbed the major hill and suddenly a much more majestic bird rose from the summit to reveal itself as a golden eagle. This unexpected sight was explained by Lewis telling us it was the survivor of a pair released on Skokholm several years earlier.

Rats had destroyed the former colonies of puffins and shear-waters, but other birds were plentiful in great variety. The most pleasing sight along one stretch of cliffs held us for hours as crimson-billed choughs battled with jackdaws for the best nesting ledges amongst its drapery of climbing plants and valerian.

Across three miles of the deepest of blue sea were views of a string of small islets belonging to Ramsey called the Bishop and his Clerks. All were unoccupied except for lighthousemen on the Bishop. Brief landings showed them to be well covered with a thin bladed grass, white sea campion, lilac mallow and sea pink amongst which the gulls and shags were nesting.

Numerous caves at the base of Ramsey's high cliffs were used as nurseries by hundreds of Great Atlantic seals many of which basked like sun-bathers on the rocky beaches. Serried ranks of nesting guillemots and razorbills groaned on their ledges just below us as we peered down in an attempt to distinguish the grey seals from their much larger cousins.

The pleasant old farmhouse was thought to be built on the site of one of the earliest saint's chapels, earlier even than the remnants of St. Justinian's chapel (A.D. 500) hidden in brambles at the cliff edge. For many centuries Ramsey belonged to the Bishop of St. David's and ancient stones and crosses marked the position of graves. In 1326 the Bishop reaped revenue from 10 horses, 100 head of cattle and 500 sheep as well as from the sale of rabbits and loads of rushes and heather. Lewis had had dreams of farming on this scale but in spite of being near the end of his resources, he cheerfully told us that although the kettle was up to its neck with hot water it still continued to sing.

A remarkable conformity with the spider and fly relationship marked our arrival at Skokholm in 1934. The boat by which we arrived in quest of spiders removed F. W. Edwards, the fly expert at the British Museum! As a matter of fact we were old friends so it was disappointing to see him go and not to have him present at the discovery of the first addition for forty years to the list of British Harvestmen or daddy-long-legs spiders (*Nelima silvatica*).

Lockley had friends staying in the house so my wife and I had come equipped with a tent, sleeping-bag and other camping needs. The first night was a disturbed one. The cocklollies or cuckles (shearwaters) came screaming in from the sea to keep us awake and then, during the early hours of the morning, a fierce storm developed which shredded the tent and blew it from over our heads. Daylight proved that the tent was beyond repair so we camped in Lockley's sitting-room for the rest of our stay.

Although we had both read of Lockley's adventures on Skokholm it was fascinating to listen to his first-hand account of his romantic arrival with his bride. In company with a carpenter he had arrived there to rebuild the ramshackle ruin of a house when a miracle occurred. An old-fashioned wooden schooner called the *Alice Williams* was wrecked in the night and promptly abandoned without loss of life. This gift from the sea provided planks, ship's timbers and a cargo of coal. Besides these treasures there were angle irons for a slipway and a brass bound ship's wheel, bells and compass for decorations. A prominent position also had to be found for the wooden figurehead of Alice Williams herself with her staring blue eyes, coiled black hair, tight bodice and smock.

Lockley's reputation as a bird expert was waxing every year and the return to Skokholm of ringed shearwaters from the distant Mediterranean where he had released them was adding to the knowledge of birds' sense of direction. His sheep farming, fishing and bird watching kept him blissfully happy and his wife supplemented his skills with her study of the wild flowers and excellent paintings. The problem of the effects of isolation on their little daughter of four was already looming in their minds and it was illustrated by watching her with their friends' children. These must have been about the first children Ann had met and she obviously found them strange. She had learned to play by herself, but not to share with others so the beginning of their island hazards was clearly on its way.

Ninth-century viking adventurers in these waters left no history, except in some of the names like Skokholm and Skomer,

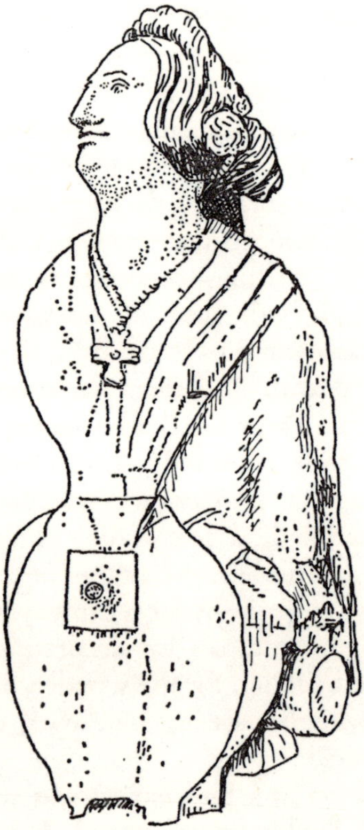

Figurehead of Alice Williams on Skokholm

but it is known that both these islands were possessed for centuries by the Earls of Pembroke, from the early 13th century onwards.

A house existed on each where a farmer lived during the summer months while sheep and cattle were being fattened for the mainland market. In winter his place was taken by ferreters who netted 3,120 rabbits during twenty-one weeks' residence on Skokholm in 1387/8. Rabbits largely fed both the ferreters and their ferrets, leaving 3,120 skins and 2,318 salted carcases for sale at Haverfordwest for a price of £13 12*s*. To this, and the profit from sheep and cattle, was added the sale of sea birds for their flesh and feathers.

One unfortunate owner had been Sir John Perrott, natural son of Henry VIII, who was clapped in the Tower of London, where he died in 1592, for impertinence to Queen Elizabeth.

Skokholm was blessed with a red soil and a neater, flatter plateau with meadows and areas of heather. It too had its share of bluebells, primroses, cowslips and orchids but somehow I preferred the greater wildness and grandeur of Ramsey and Skomer.

Far out to sea was the tiny green mound known as Grassholm where gannets have their only nesting colony in the whole of England and Wales. This had been visited with Lockley in 1930 but it was two later visits in 1959 which left their greater mark.

My cousin Bobbie emulated Nelson in believing Milford Haven to be the best natural deep water harbour in Britain, so he was having a documentary film made in which he wanted to include pictures of this near-by bird island. I now found myself in Milford Haven as guide and commentator along with the film producer and his team of photographers and sound recorders. This novel experience was to be full of surprises and temperament.

To my horror it was found that scene one was to be enacted on the quayside where the *great* naturalist was to be welcomed by the producer in stilted sentences reminiscent of 'Dr. Livingstone I presume'. By the time I had converted this script into innocuous greetings the producer had disappeared round the corner where he was found making himself up in front of a looking glass and fitting plumpers in his mouth to remove sagging cheeks.

The sun disappeared behind clouds just as he had finished these operations and a plea to scrap the scene was finally accepted after a delay of nearly an hour.

The next scene of a general talk on what we were hoping to see was to be taken in mid ocean. This was rehearsed and taken with the new boy trying to forget he was being photographed while the producer seemed to be remembering his earlier stage experience. This he had the grace to admit, so the scene was retaken. Now he had anxieties lest the sound and vibration of the engine had spoiled the result, so it was taken again with the engine mute. This time there had naturally been no wash from the boat so the scene was taken for the fourth time after starting the engine and filming directly it stopped. A child of ten could have told him that the surge of white waters would fade directly the propeller ceased to revolve.

Panic set in when it was mildly pointed out that most of the morning had passed and that twelve miles of sea lay between us and Grassholm. The suggestion that we should compromise for Skokholm instead brought so sharp a rebuke from the guide/commentator that we proceeded in complete silence which lasted until our arrival, due partly to the Atlantic swell which can quieten many a landlubber.

The rise and fall of the sea prevented the boatman from tying up, so the producer and I jumped on to a rock with a view to prospecting ahead of the others. A few kittiwakes nesting on a ledge provoked the producer to a petulant query as to whether he had braved sea sickness for this. In order to make the shock of what he would see later on still more stupendous a course of not undeserved teasing was adopted. 'Kittiwakes,' he was told, 'are beautiful birds whose graceful flight is a particular feature of the western coasts of Britain, seen much more rarely off the eastern coast.' Further extolment fell on deaf ears as he panted up the steep grassy hill slope falling at intervals into collapsed puffin burrows.

'What is that ghastly smell of decaying fish?' he asked, as we neared the summit. 'Decaying fish,' I replied with a grin which brought no response.

And then came the moment of fantastic beauty which left his artistic soul dumbfounded. Just a few yards below the hill crest and extending as far as the eye could see in both directions were some 20,000 gannets massed side by side on their nests. In the air above us and out to sea were thousands more showing their unrivalled mastery of flight, their white bodies and black-tipped wings gleaming against the blue sky. Now and then one of these dived vertically into the sea from a height of 200 feet to catch and swallow a fish (Pl. 13).

Meanwhile these huge birds, the size of geese with a wingspan of six feet, were harshly rebuking us from their nests. Assuring the producer of his safety I drew him close to the nearest birds where we could both see their snowy plumage sprinkled with fawn around their heads, their spear-tipped beaks and baleful eyes. Closer still and they lunged fiercely at us, never quite striking our legs but occasionally showing us their single greenish blue egg or speckled chick. One or two disgorged a half-digested fish at our feet. Others in this city of birds were wagging their heads or knocking bills together in courtship.

From now on I became 'my dear chap' in every sentence, but his exaltation was of short duration. He ran down the hill slope, not caring about the occasional tumble into a puffin burrow, while I followed at a more leisurely speed in time to witness a terrible scene of frustration.

The boatman was pointing to rocks below the surface on which his boat would be thumped if he drew closer to land. The photographer and sound recordists were refusing to jump with their valuable apparatus. Leaving the argument to pursue its course, another rock cranny was found where deep water would make a landing more easy, but by the time I returned it had been decided that the afternoon was too far spent to start work at all! None could stay another night in Milford Haven so the weekend had been wasted except for me who had enjoyed it.

An arrangement was made to return to Milford Haven on the Friday two weeks later for an early departure for Grassholm on the following day. I had insisted on a Saturday visit because I had to be at my office desk on Monday morning and there were no Sunday trains from Milford Haven at all. A telephone message from the producer's secretary just before I left home told me that everything was arranged for the *Sunday*—a car to bring the team and their equipment from London, accommodation in Milford Haven and a boat. Expostulation was useless so it was made clear that somehow they would have to squeeze me into that car on the homeward journey in time to catch the last train from a station nearly 100 miles away.

Knowing something of my friends by this time a telephone call was put through to the hotel at Milford Haven only to find that rooms had been booked for everybody except the guide/commentator. Feeling mischievous rather than angry, and bearing in mind that I had volunteered to do my task for expenses only, I arranged with the friendly manager that I should oust the producer from the best room for whom a small one would be found.

The producer had sacked his cameraman, his sound recorders and his boatman, so it was a fresh team which sailed for Grassholm this time. The excellence of the resulting film made me wonder if art in this industry often springs from the ashes of inefficiency.

The dream of owning an island came nearest to reality within a year of visiting Ramsey in 1933. After thinking what I would do with it, I put in a bid for this, my favourite island.

Donne had written in 1624 that 'No man is an Iland, intire of it selfe; every man is a peace of the Continent, a part of the Maine,' and I had seen enough to realise that continuous island life away from friends would never suit my temperament. Solitude brings no happiness except in small doses and the nearness of the romantic to tragedy is close when fighting for bare existence on a small island. On the other hand, life on large islands was not worth substituting for that on the mainland.

If Ramsey were to become mine it would not, I decided, change my way of life. A search would be made for a nice tenant farmer interested in birds who would graze his sheep on the island's lush pastures free of all rent, provided part of the house would be mine for the entertainment of friends whenever we could go there.

The history and the natural history would be lovingly compiled. The house would be repaired, the harbour improved, a reliable boat would be acquired and lobster pots laid. War would be declared on the rats and a mate found for the solitary golden eagle. Precious days would be spent there dispelling the cares of the outside world.

These and other dream thoughts were fun while they lasted, but the auction price exceeded anything I could afford.

Acknowledgements

Through all the years that islands have lured me people and books have helped with their company and encouragement, yet now I can only mention a few. Tom Lethbridge more than any other, my wife Helen, my daughters Ginie and Richenda, Richard Lynex, Eric Classey and many others. Unfortunately, Maxwell Knight is no longer alive, nor Sir James Wordie, nor Sir Stephen Courtauld.

I am grateful to Sandra Raphael for corrections and to Nial Cossart for allowing me to see and use his family scrapbook on Madeira and the Desertas.

I am grateful to Sandra Raphael for corrections and for compiling the Index. Thanks are due to many others, including Gerry Maul and Alec Zino in Madeira, to Dr. W. H. Wince and Peter Brett for photographs, and T. C. Lethbridge and A. Langham for drawings.

A short list of books follows:

Anderson's *Guide to Orkney.*

Bernadus, Fr. Maria *Cistercians de Lerins.*

Boswell, James *A Tour of the Hebrides.*

Bristowe, W. S. 'The Fauna of the Arctic Island of Jan Mayen'. *Ann. Mag. Nat. Hist.*, XV, 1925.

Bristowe, W. S. 'A Preliminary Note on the Spiders of Krakatoa'. *Proc. Zool. Soc.*, 1931.

Bristowe, W. S. 'The Liphistiid Spiders.' *Proc. Zool. Soc.*, 1932.

Bristowe, W. S. 'Hunt for a Wolf Spider.' *Animals.* Sept. 1963.

CAMPBELL, J. L. *'Gaelic Folksongs from the Isle of Barra'* Linguaphone Institute.

CLIFFORD, SIR HUGH *Heroes in Exile.* Smith Elder, 1906.

COLLIS, MAURICE *Raffles.* Faber and Faber, 1966.

DASENT, SIR G. W. *Icelandic Sagas,* 1894.

DAMMERMAN, K. W. *The Fauna of Karakatau.* N. V. Noord-Hollandsche Uitgevers, Amsterdam, 1948.

DE LA MARE, WALTER *Desert Islands.* Faber and Faber, 1930.

DUFFERIN, LORD *Letters from High Altitudes.* John Murray 1858.

FURNEAUX, R. *Krakatoa.* Secker and Warburg, 1965.

GARDNER, E. A. *Greece and the Aegean.* Harrap, 1933.

GIRALDUS CAMBRENSIS Bohn's Antiquarian Library, 1881.

GORDON, SETON *The Charm of Skye.*

KENDRICK, T. D. *A History of the Vikings* Methuen, 1930.

LANGHAM, A. and F. *Lundy, Bristol Channel.* Broadacre Books, 1959.

LETHBRIDGE, T. C. 'Excavations at Kilpheder, South Uist.' *Proc. Prehistoric Soc.,* 1952.

LOCKLEY, R. M. *Dream Island.* Witherby, 1930.

LOCKLEY, R. M. *Island Days.* Witherby, 1934.

LOCKLEY, R. M. *I know an Island.* Harrap, 1938.

LOYD, L. R. W. *Lundy, Its History and Natural History.* Longmans, 1925.

MACLEOD, KENNETH *The Road to the Isles.* A. and C. Black.

MARQUET, JEAN *Marie Ier Roi des Sedangs.* Hué, 1927.

MARTIN, MARTIN *A Description of the Western Islands of Scotland.* 1695.

MERCANTON, P. L. 'La Première Ascension du Beerenberg de la Jan Mayen.' *L'Echo des Alpes,* 1924.

MESAROVIC, J. *Korcula.* Belgrade.

MITTON, G. E. Article on Sighiri in *Wonders of the Past.* Hammerton.

Osterreichische Polarstation Jan Mayen, 1882–83. Vienna, 1886.

PERRY, R. *Lundy, Isle of Puffins.* Lindsay Drummond, 1940.

ROBSON, R. W. *Pacific Islands Year Book.* Pacific Publications, Sydney, 8th Ed. 1959.

ROUGHLEY, T. C. *Wonders of the Barrier Reef.* Angus and Robertson, 1956.

ROYAL SOCIETY *The Eruption of Krakatoa.* 1888.

SOLJAN, ANTON *The Thousand Islands of the Adriatic.* Belgrade.

SOULIÉ, MAURICE *Marie Ier Roi des Sedangs.* Marpan, 1927.

STEHN, C. E. *Krakatau.* 4th Pacific Science Congress, 1929.

STURLASON, SNORRI *Heimskringla.* Methuen, 1932.

SWINSON, A. *Scotch on the Rocks.* P. Davies, 1963.

SWIRE, O. F. *The Outer Hebrides and their Legends.* Oliver and Boyd, 1966.

WORDIE, J. M. 'Jan Mayen Island.' *Journal Royal Geographical Soc.*, 1922.

INDEX